T0176746

Communication Practices in Engineering, Manufacturing, and Research for Food and Water Safety

Communication Practices in Engineering, Manufacturing, and Research for Food and Water Safety

Edited by

David Wright
Missouri University of Science and Technology

IEEE PCS Professional Engineering Communication Series

IEEE PRESS

Library of Congress Cataloging-in-Publication Data is available.

ISBN: 978-1-118-27427-9

Printed in the United States of America.

Contents

6 Influences of Technical Documentation and Its Translation on Efficiency and Customer Satisfaction   **145**

Elena Sperandio

7 Communicating Food Through Muckraking: Ethics, Food Engineering, and Culinary Realism   **171**

Kathryn C. Dolan

A Note from the Series Editor

"Like most humans, I am hungry ... our three basic needs, for food and security and love, are so mixed and mingled and entwined that we cannot straightly think of one without the others."
—M.F.K. Fisher, *The Gastronomical Me*

The IEEE Professional Communication Society (PCS), with Wiley-IEEE Press, continues its book series titled *Professional Engineering Communication* with this collection curated by Dr. David Wright. This book, *Communication Practices in Engineering, Manufacturing, and Research for Food and Water Safety*, brings together the thoughtful research and perspectives from professionals in different fields, all writing to the ways in which communication efforts affect the ways we think about food and water and the ways in which it arrives at our doorstep. The rhetorical frameworks that provide scaffolding for personal opinions, public policy, research and development, and advertising are intricate and fraught with human emotion (both good and bad). The power of these chapters is how the reach into the dialogue and deconstruct the communication's underpinnings.

From a larger perspective, this book is a welcome addition to the *Professional Engineering Communication* (PEC) book series, which has a mandate to explore areas of communication practices and application as applied to the engineering, technical, and scientific professions. Including the realms of business, governmental agencies, academia, and other areas, this series will develop perspectives about the state of communication issues and potential solutions when at all possible.

The books in the PEC series keep a steady eye on the applicable while acknowledging the contributions that analysis, research, and theory can provide to these efforts. Active synthesis between onsite realities and research will come together in the pages of this book as well as other books to come. There is a strong commitment from PCS, IEEE, and Wiley to produce a set of information and resources that can be carried directly into engineering firms, technology organizations, and academia alike.

At the core of engineering, science, and technical work is problem solving and discovery. These tasks require, at all levels, talented and agile communication practices. We need to effectively gather, vet, analyze, synthesize, control, and produce communication pieces in order for any meaningful work to get done. It is unfortunate that many technical professionals have been led to believe that they are not effective communicators, for this only fosters a culture that relegates professional communication practices as somehow secondary to other work. Indeed, I have found that many engineers and scientists are fantastic communicators because they are passionate about their work and their ideas. This series, planted firmly in the technical fields, aims to demystify communication strategies so that engineering, scientific, and technical advancements can happen more smoothly and with more predictable and positive results.

Traci Nathans-Kelly

Preface

This book is a collection of perspectives on both the history and the current state of food and water engineering. More specifically, it represents some of the many ways that our food and water quality are affected by communication.

Despite our many technological advances, contaminated food and water continue to take lives worldwide. We tend to think of these events as happening in some far away, underdeveloped portion of the world. While it is true that countries with poor infrastructure are subject to more incidences of contaminated food and water, we continue to suffer the same types of problems in the United States and Europe. One does not have to look far to find stories of lives being claimed by *Escherichia coli* in Washington State or outbreaks of food poisoning at the 2014 Food Safety Summit in Baltimore, Maryland.

Technology certainly helps to lessen the impact of those outbreaks, and to track outbreaks of foodborne illnesses to their source more quickly. Similarly, stories from the past serve as good reminders of previous failures. But technologies must be accepted by food producers and the public to be of value, and lessons from the past are continually mitigated by the need for profits and development. And in some cases, such as with genetically modified foods, the very technologies we have created spur new debates about food safety and ethics.

Therefore, communication continues to play a vital role within food production industries as we struggle to implement new technologies within industry while maintaining policies that protect consumers. We hope that this book will help readers better understand where we come from, where we are, and what is being done to improve food safety through communication.

David Wright

List of Contributors

Becca Cammack has worked in the environmental and energy industries since graduating with her B.S. in Geology from Northern Arizona University in 2000. After spending five years working in various roles and disciplines in the consulting industry, she joined the southern California energy industry as an environmental professional specializing in water quality permissions and compliance. She presently works as Senior Environmental Scientist with Pacific Gas & Electric, Co., in northern California, where she supports project teams in their efforts to understand and comply with environmental permits and regulations. She has obtained a professional certificate in Technical & Scientific Writing and is completing coursework toward a Master of Arts degree in Rhetoric & Writing Studies, both from San Diego State University.

Roy E. Costa is Registered Sanitarian (RS) and President of Environmental Health Associates. He has more than 33 years of environmental health practice in the academic, government, and private sectors. Mr. Costa is a food safety consultant, educator, auditor, and expert with international experience. Mr. Costa is the author of recognized Internet food safety courses and classroom-training programs and numerous publications in the area of food safety.

Kathryn Cornell Dolan, Ph.D., is Assistant Professor at the Missouri University of Science and Technology, Rolla, Missouri, USA, where she teaches early US literature with a focus on food studies, globalization, and environmental criticism. Her book *Beyond the Fruited Plain: Food and Agriculture in U.S. Literature, 1850–1905* is available through University of Nebraska Press.

William K. Hallman, Ph.D., is Professor and Chair of the Department of Human Ecology at Rutgers University, New Brunswick, New Jersey, USA, and serves as the current Chair of the Risk Communication Advisory Committee of the US Food and Drug Administration. An expert in risk perception and risk communication, his research focuses on food safety, food security, and public perceptions of controversial issues concerning food, technology, health, and the environment.

Edward A. Malone, Ph.D., is Professor of Technical Communication at Missouri University of Science and Technology, Rolla, Missouri, USA. He has published articles in the *Journal of Business and Technical Communication, IEEE Transactions* *on Professional Communication, Technical Communication, Technical Communication Quarterly*, and most recently *Journal of Technical Writing and Communication*.

Havva Malone (née Tezcan) has a Master of Science in Physics and is a part-time teacher in Rolla, Missouri, USA. She started the Rolla School District's first FLL team, Global Dreamers, and served as the team's coach. She is currently working on a book that communicates math concepts to primary and secondary students through photographs.

Mary L. Nucci, Ph.D., is Research Assistant Professor in the Department of Human Ecology at Rutgers University, New Brunswick, New Jersey, USA. Her research focuses on the public perceptions and communication of science in media and informal education.

Elena Sperandio lives in Berlin, Germany. After completing degrees in German and Romance studies and traffic engineering at the Technische Universität, Berlin, and Freie Universität, Berlin, she cofounded 4-Text Global Translation & Localization Services, where she has since gathered more than a decade of experience in project management, translation memory systems, and quality assurance. She has detailed knowledge of the problems faced by large companies in their struggle to provide good, consistent technical documentation. In 2009, she became the CEO of 4-Text.

David Wright, Ph.D., is Associate Professor at Missouri University of Science and Technology, Rolla, Missouri, USA, where he codirects the technical communication program. He studies technology diffusion, technical communication history, and international issues related to agricultural development. He has published articles in *Technical Communication Quarterly* and *Journal of Technical Writing and Communication*.

Acknowledgments

This book has been a challenge on several fronts. At times I wondered if it would ever be completed. We tend to forget that life happens regardless of whether we have a project in the works. Special thanks goes to Traci-Nathans Kelly for helping to keep this project alive through those difficult times and for helping to make this series a reality. She has been an excellent guide for our team and an even keel throughout.

We also wish to thank Mary Hatcher, Kenneth Moore, and the entire team at Wiley-IEEE Press for their help in the publishing process.

Finally, on behalf of all of the authors, I thank the many family members and colleagues that invariably become laborers in the publishing process. We are grateful that our thanks are enough for them.

Cowboys and Computers: Communicating National Animal Identification in the Beef Industry

David Wright

1.1 Industries Collide

In May of 2005, I began working with a small software firm that develops software that would allow data ownership and transfer on a granular level. The idea was that rather than transferring entire documents or entire records, small pieces of information could be shared on the basis of permissions and commerce. To the firm, it seemed like a fine idea and a practical one. As I learned more, I discovered that the software was being specifically developed in response to a push by the United States Department of Agriculture (USDA) for a National Animal Identification System (NAIS) capable of tracking cattle moving through the supply chain. In theory, new technology and new methods of doing business would speed commerce and, more importantly, allow for swift containment of any disease outbreaks that might threaten consumers.

Upon being assigned to the project, my first thought was that this seemed like a significant advancement for the industry. As the technical communicator on staff, my first task was to establish relationships with potential customers and begin to gather information on what tools they would like to see included in a software product. Imagine my surprise, then, when one of my first cattle auction contacts told me that he did not understand the NAIS and did not need to.

Communication Practices in Engineering, Manufacturing, and Research for Food and Water Safety, First Edition.
Edited by David Wright.
© 2015 The Institute of Electrical and Electronics Engineers, Inc. Published 2015 by John Wiley & Sons, Inc.

In his opinion, the NAIS was nothing more than an elaborate plot orchestrated by the Internal Revenue Service to spy on cattle producers throughout the country.

This individual was successful, widely respected, and by no means alone in his opinion of the NAIS as a waste of time and money. His ideas about the true motives for the NAIS were perplexing, to say the least.

Although the original NAIS plan was scrapped in 2010, that was not the end of the story. In April 2013, the USDA launched a new program designed to be much less restrictive and transparent to members of the beef industry; it would rely much less on the use of computer technology in particular. Whether the new plan will enjoy more widespread acceptance remains to be seen, but the USDA's retreat from the earlier initiative shows us that diffusing technology into an industry like the beef industry is not simply a matter of producing that technology but an exercise in communicating new technology to members of that industry.

The story of the original NAIS plan's ultimate failure within the beef industry is also a warning to other industries and agriculture in general, in that it illustrates potential difficulties when it comes to implementing widespread technologies. For technical communicators, this means that we should be asking ourselves very pointed questions now:

- What do we need to know about this situation (and similar future situations) in order to predict successful communications in the future?
- How can we best study industries and their technology?
- How can technology diffusion be successfully enhanced through research and targeted communications for specific audiences?
- When conflicts between industries do arise, what is the best way to ensure successful communication?

The story of the NAIS and the beef industry offers a preview of situations to come in which entire industries are resistant to new information technologies. Unfortunately, investigating the communicative failures that take place within supply chains is complex. Communications are not limited by form. They come in paper and digital forms, over cell phones, through policy statements from governmental agencies, and from industry alliances that wield great power over industry opinion. Finally, they come through the elusive art of personal communication. Stakeholders in some traditional industries are often not available in chat rooms or on our e-mail servers or through any of the other high-tech means of communication that have segregated us from each other.

This, however, does not mean that they do not communicate. These industries have established, complex communication networks that often play vital roles in technological diffusion. Technical communication must, therefore, develop methods that take into account the various forms of communication that are now available while maintaining an ethnographic perspective and methodology that investigates industries at ground level. We need holistic approaches to understanding intricate problems. This chapter is one attempt to do so and to provide insight for the future. It also offers a set of tools for research that may give future technological deployments more success. Perhaps these insights will be useful to future practitioners and academics alike because of their direct link to the new technologies that show no signs of slowing down and continue to affect our lives at an increasing pace.

1.1.1 Resistance to Technology in the Beef Industry

The beef industry is steeped in tradition. It is an industry that does not change rapidly and does not readily adapt to change without good reason. Moreover, sometimes this industry has been unwilling to adapt to new technologies even when there would seem to be very good reason for it. Some sections of agriculture (and I will focus just on the beef industry in this chapter) have been slower to adopt new ideas and information technology. There are many factors that go into the lack of technological diffusion seen in the beef industry, not least of which is the fact that industry leaders see no need for new information technologies.

Yet, there have been substantial advances in other facets of the beef industry during recent years. For example, veterinary medicine has made remarkable advances in the treatment and prevention of many diseases and common animal sicknesses. Nasal sprays, injections, and other treatments enjoy widespread use to prevent animal deaths during transport and growth. Likewise, new techniques that include product branding, prepackaging, and efforts to make beef products healthier have radically changed supply chain management and distribution.

Yet, in one technological aspect, the beef industry has been at a standstill. I speak of computer and software technology in general and the slow pace at which the beef industry as a whole has adopted new methods of doing electronic business, even in the face of potentially disastrous consequences. While the beef industry seems perfectly willing to accept some types of innovations, computer technology and animal identification have been shunned, at least by many segments of the beef supply chain.

1.1.2 Having a Cow Over Mad Cow Disease

Still, the resistance to technology would be more understandable if the matters at hand were those of general office automation designed to make daily tasks easier. But beginning in December 2003, when the first case of "mad cow disease" or BSE (*bovine spongiform encephalitis*) was discovered in the United States, technology became a much more serious concern for the beef industry. In fact, the market for US beef changed so much that beef exports fell from more than US $3 billion in 2003 to barely US $500 million in 2004 [1]. The scare over the outbreak of mad cow disease led numerous countries to close their borders to US beef, and the fears hurt domestic sales as well.

The debate over using technology in the beef industry began in earnest at that time. Consumers were concerned about BSE because of the deadly effects of infected beef on humans. Adding fuel to the fire was a recent episode of BSE contamination, located in Alabama; USDA and state officials investigated five auctions and 36 different farms with DNA testing equipment in an unsuccessful effort to locate the source of contamination

[1]. Worldwide, more than 150 deaths had been linked to infected beef from numerous sources, and there may have been many other misdiagnosed cases in underdeveloped countries. Cattle producers were also concerned. Similar incidents including outbreaks of foot-and-mouth disease have led to the slaughter of hundreds of thousands of cattle, sheep, and pigs in numerous countries (most notably the United Kingdom in 2001) because of inability to trace the disease to its source or contain the outbreak [2].

After the 2003 BSE outbreak, Japan cut off shipments of US beef, as did many other countries including Canada and South Korea. The Japanese market was significant, as was the Canadian market. Japan alone bought US $1.3 billion worth of US beef in 2002, but then promptly cut off imports after the outbreak. Likewise, South Korea, which imported US $815 million worth of US beef, cut imports to zero. Although in 2006 both countries resumed imports after thorough inspections of US processing plants, they were still very wary of US ability to control and trace disease, and consumer/government confidence in those countries has yet to return to normal as of 2014. Thailand, China, South Korea, and Singapore, all significant importers as well, still had bans on US beef as of 2006 [3]. While South Korea and Japan have relaxed their stance on US beef since then, China (a major market) still bans US beef. The other countries still place restrictions on those imports [4].

In the wake of the scandal, the USDA and beef industry professionals began to seek answers to both national and international concerns. But industry resistance to a mandatory plan, coupled with a disagreement among industry associations, alliances, and businesses, crippled the proposed program and brought it to a standstill.

In addition to natural disease concerns, the fear of agriterrorism, which might produce an introduced disease, became a real concern at about the same time. After the terrorist attacks of September 11, concerns led many citizens and government employees alike to question the ability of terrorists to sabotage the US food system through biological means. Animal illnesses can spread rapidly and be difficult to trace [5].

In response, the USDA launched an initiative to implement a national animal tracking system that would allow animals to be tracked and traced to their point of origin in case of a disease outbreak. On paper, the idea was fairly simple. Animals would be tagged with an electronic radio frequency tag that contains a unique 15-digit identifier. This tag would then be scanned into computer software that would store the number, along with other vital statistics (perhaps), in a database. Then, as the animal moved through the supply chain, database administrators could track the animal on the basis of that number by continually updating the animal's location on the basis of new scanning information (presumably the animal would be scanned at each new location). This strategy should have, in turn, allowed for swift containment of any disease or potential harm to consumers and boosted both domestic and international confidence in US beef, because any animal's location could be easily tracked and any animal's place of origin could be known almost immediately.

Typically, after being raised on the farm for a given period of time, an animal moves from its place of birth to a livestock market, where it is sold to either an order buyer or directly to a packing company. (Note: many animals also move directly from the farm to a stocker or a packing company but most do not.) Figure 1.1 shows a typical beef supply chain. If the animal is purchased by an order buyer, it typically moves to either a

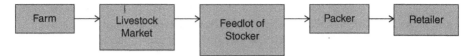

FIGURE 1.1. A simple beef supply chain from calf birth to harvest.

stocker operation, where it will feed on grass, or to a feed lot, where it will be fed grain. In either case, the animal eventually moves on to a packer for processing and then to a retailer like Walmart or McDonald's.

It would seem from this simple explanation of animal movement that tracking cattle from birth to our dinner tables should be quite manageable. Unfortunately, an animal's actual movement is often much more complex. Because of market influences that affect the way cattle are bought and sold, many animals move from one level of this supply chain to another and then back and forth again several times before finally arriving at our local retailer. It is not uncommon for a group of calves to be brought to a livestock market for sale together, having just left their place of birth. But at that point they may be sold separately to different buyers, sent to multiple locations, and then resold from those locations or at different auctions. In fact, many animals are bought, sold, and moved multiple times during their lives. They may spend that time in multiple states and with multiple owners. The overall effect of these transactions is a movement pattern that is very difficult, time consuming, and expensive to trace.

Still, societal health and safety concerns prompted the USDA and some beef industry companies to begin to look for a response to national and international pressure. As this technological niche within the beef industry began to open, companies specializing in software that would allow animal tracking and data collection began to spring up. In addition to health-related data, software companies sought ways to integrate individual animal data with supply chain data while assigning unique identification numbers to those animals. Their goal was to introduce new software and technology to the beef industry. If they were successful, then animals and financial data could be moved and tracked together, offering safety and possibly financial rewards.

But the sheer size of the problems associated with national animal identification meant that computerization would be essential for the beef industry to move forward in the areas of animal identification and electronic commerce. There are approximately 90–110 million beef cattle in the United States at any given time [6]. Of these, packing companies harvest between 35 and 40 million annually.

To deal with the high numbers of animal turnover in combination with animal identification, the USDA began the development of the NAIS in 2003. The proposed system would integrate three components: premises identification, animal identification, and animal tracking.

Premises identification was to be done only once per site, with a premise identification number assigned to each homestead or ranch where animals were raised. This part of the plan was a relatively uncomplicated process. However, animal identification and animal tracking would have required significant upgrades in computer technology within the beef industry. The USDA proposed a system requiring a 15-digit unique identifier for each animal in the country. This number would be assigned to the animal on the premises where it was born, then recorded and tracked within a centralized database (or, as we will see later, perhaps numerous databases) as the animal moved through the supply chain. The animal could then be traced to its origin in the event of a disease outbreak.

Tracking and tracing the national herd, then, would have required approximately 100 million unique 15-digit numbers at any given time, each assigned to an animal in the national herd. Then, 35–40 million new animals would need to be assigned these numbers on an annual basis at birth. In addition, another 35–40 million unique identification numbers would need to be tracked at any given time as they moved through the supply chain from producer to livestock auction to feeder to stocker and ultimately to the packer, where they are harvested and processed. Clearly, this task could not be accomplished without the aid of computer technology.

There was certainly no lack of opportunity for the beef industry to adopt this new technology. In 2006, *Drovers* magazine listed over 50 companies in their directory of service providers for animal identification products [7]. Most of these companies offered software and/or database services to support the NAIS program. As Bob Scheiler of the *Wall Street Journal* reported, "The effort has sparked something of a stampede among makers of radio identification tags, retinal scanners, and other electronic gizmos with the potential to keep tabs on millions of animals from birth to slaughterhouse" [8, p. B.3].

In addition, many of the top US retailers like McDonald's®, Walmart®, and Tyson® have actively encouraged the beef industry to adopt technology and a system that allows for animal traceability. Robert Cannell, McDonald's United States Supply Chain Director, publicly stated his desire for animal identification on numerous occasions. In an issue of *Farm Week*, he said that while McDonald's was sensitive to the fact that some beef producers needed both time and money to implement the NAIS, McDonald's would not wait for the USDA to act. According to Cannell, McDonald's would actively seek contracts with beef suppliers who could track all animals [9]. Because McDonald's traditionally buys over 1 billion pounds of domestic beef annually, its desire for traceable beef would seem a strong incentive for members of the beef supply chain to act in developing the technology needed to satisfy this request. Yet, even now in 2014, while McDonald's is able to trace 100% of its beef in countries like England and Ireland, the company is unable to do so in the United States.

1.1.3 Change Is Slow in the Beef Industry

Information technology within the beef industry has always been a hit-or-miss affair. Some beef professionals tend to be early adopters of new technology, while others refuse to adapt to new technologies even in the face of obvious benefits. The beef industry has been slower than most to adopt new technology for various reasons including expense, low return on investment, and a general lack of awareness concerning new technologies and their benefits. Even in 2006, it was not uncommon at livestock auctions to find dot-matrix printers, DOS-based computer programs, and hardware that would be considered ancient by most standards. While some offices were modern, most were not. The fact that some of the hardware I have seen in livestock markets still functions at all is a testament to manufacturers. It is worth noting that this does not make their owners "backward" or behind the times. It simply means that current computer technology is not necessary for profitability and the usefulness of newer technologies is suspect. Since that time, many of those auctions have upgraded their hardware, but not necessarily their software as it relates to their business operations.

Secretary of Agriculture Ann M. Veneman asked the USDA's Chief Information Officer to make creating a national animal ID system his top priority in 2003 [10]. Afterward, the date for the system's debut changed repeatedly and eventually no longer included mandatory participation. Those who favored a national plan differed sharply on how that plan should be designed and administered, and many within the industry believed that the system will never mature at all.

Yet after Veneman's 2003 mandate, the USDA pushed forward with its technical requirements. In March 2006, the USDA released performance standards for animal identification tags that included life span, reading distance, and failure rate specifications [11]. At roughly the same time, the USDA announced plans to allow numerous private databases to comprise a national system of identification and traceability. And in August 2006, the USDA approved a Minnesota company as the first manufacturer of "approved" NAIS RFID (radio frequency identification) tags.

However, private industry groups including many producers and politicians, like Representative Ron Paul of Texas, organized to block the system entirely except in the case of international sales. They saw the NAIS as unnecessary government interference and an infringement upon personal liberties. Articles continually appeared by authors like Jodie Gilmore, who stated that "The USDA has targeted farms and livestock facilities and their livestock for intrusive, unnecessary, and eventually mandatory identification and tracking regulations" [12, p. 1].

1.1.4 Communication Breakdowns and Coffee Shop Policymaking

In fact, as I will show, miscommunication and poor communication directly contributed to the confrontational atmosphere that developed surrounding the debate over the NAIS. While we can undoubtedly place part of the blame for the lack of technological development within the beef industry upon cultural resistance to technology and economic issues, it is also partly because of (i) miscommunication and poor communication by national entities, and (ii) misunderstanding promoted by communication networks about the tracking initiatives themselves.

To understand some of this story, it needs to be said that the beef industry is a very personal business, and communication circles are tight. There are defined networks of individuals who share information and opinions to reach a consensus. This process is not unlike what Rogers and Kincaid describe when they write, "It seems that almost everyone depends heavily on interpersonal communication channels to obtain the information that he/she needs to make important decisions" [13, p. 344]. In other words, we turn to those we know personally and trust for information about difficult decisions.

While the basic message coming from portions of the supply chain (some large packing companies and retailers) and the USDA seemed clear (that animal tracking must be implemented), the specifics of that message needed to persuade others did not reach the beginning of the supply chain (small producers and livestock auction owners). If those important messages had been delivered well to small producers and auction owners, there could have been much wider understanding of the logistics of the system and of the computer and technology required by the proposed NAIS. Those opposed to the NAIS might not have accepted it still, but they would have known much more about what, exactly, they were rejecting.

Much of the resistance to the proposed NAIS plan can be traced to communications concerning animal identification and technology. Although the USDA actively promoted national animal identification, communications about the business model and technologies that would necessarily accompany such a system were slow to change resistance, especially among livestock auction owners and producers. Part of the USDA's problem in communicating directly with constituents at "ground level" was the influence of industry alliances—those organizations that are voluntarily joined by beef industry professionals.

> Those alliances are much better at keeping in touch with their membership than the USDA and are generally viewed by independent operators as much more in touch with the needs of common beef industry professionals than the government or hi-tech firms.

The sole mission of these alliances is to protect and create profit for their members. Membership in the alliances is tightly monitored, and information flows through their members to other members in the form of mailings, meetings, and informal, personal communication.

My experience working in the industry showed me that informal personal communication is still the predominant form of truly persuasive communication in the beef industry. Like it or not, more opinions are formed in the coffee shops of livestock auction markets and local cafés than are formed anywhere else. Part of this is because of the close-knit communities that make up the beef industry.

But another, perhaps more influential reason is that opinions in the beef industry rely upon established communication networks that are often far removed from mass communication. I have yet to hear a cattle owner in a coffee shop say, "Do you know what I read on the internet this morning?" But the influence of interest groups and alliances goes far beyond common knowledge. Members of alliances, including producers, market owners, buyers, and packers, believe that the alliances are their best (and perhaps the only) way to control their collective destinies through congressional lobbying and solidarity as an industry.

However, the various industry alliances have had difficulty agreeing among themselves on plans for new technology and animal identification, and for that matter, on many different aspects of the beef industry. For example, the National Cattleman's Beef Association (NCBA), which represents members of the beef supply chain including producers, feeders, and processors, and the Livestock Marketing Association (LMA), which is the organization that represents livestock market (cattle auction) owners, have historically agreed on very little concerning the NAIS. Both developed their own plans for national animal identification in response to the USDA's original push. Those plans differed significantly from each other, and also from plans mapped out by the USDA. For example, the LMA requested that the federal government fund tagging services at auctions and manage the corresponding information database, while the NCBA favored entirely private tagging and database services. Then there were other producer organizations, such as the Ranchers-Cattleman Action Legal Fund (R-Calf), which is centered in northern states, and also had its own agenda. Taken together, these differing points of view offered little consensus within the industry. When feedlots (where animals are typically "fattened up"), packers (who process the meat), and the various large-scale national retailers (Walmart, McDonald's, etc.) were added to the equation (each of whom had their own agendas), wide scale confusion and disagreement ensued.

Although there have been many articles, and even more opinion pieces, written in recent years about the debate between those for and against the NAIS system, there have been almost no attempts to analyze the lack of understanding that existed between those who supported a national animal identification system and those who opposed one.

1.1.5 Can We All Just Get Along?

Meanwhile, the USDA, seemingly paralyzed by this lack of consensus, struggled to come up with a plan that would suit all parties. A general uproar ensued, as all parties involved vied for control of both the proposed system and the data stored within that system. In fact, the potential benefits of animal identification were so misunderstood and so poorly communicated to members of the supply chain that they went largely unnoticed. For example, Representative Ron Paul [14] of Texas issued a statement through the US Federal News Service in which he called the National Animal Identification System nothing more than a tax on livestock owners that would allow the government access to detailed information about their private property.

The *Countryside and Small Stock Journal* (and other small journals as well) routinely printed critical articles concerning the NAIS. For example, Zanoni [15, p. 1] says that the NAIS will "[...] drive small producers out of the market, will prevent people from raising animals for their own food, will invade Americans' personal privacy, and will violate the religious freedom of Americans whose beliefs make it impossible for them to comply." Her assertions are debatable, but my investigations show that many producers agreed with her. They often cite (now, still, just as in 2006) government interference as one of the primary reasons for their resistance.

Taking things a step further, Senator Jim Talent of Missouri, who is also a member of the Senate Agriculture Committee, coauthored a bill in 2006 that would have prohibited the USDA from implementing a mandatory National Animal Identification System. Talent [16, p. 1] said, "The development and implementation of an animal identification system must be voluntary and done with the cooperation of producers rather than by coercing them to participate. I have made this point repeatedly to the USDA." Surprisingly, opponents of the NAIS said that the bill did not go far enough to crush animal identification and did not support the bill.

While the debate raged, the beef industry continued to search for a system that would alleviate the fears of both US and international consumers. At the same time, the USDA seemed to be losing its ability to make any decisions about national animal ID, which only added to the confusion. That lack of authority was especially true among "ground-level" members of the beef industry. The problem was exacerbated by the exclusion of a major component of the beef supply chain. Livestock markets, which form a large part of the beef supply chain, were excluded, for the most part, from programs like the Process Verified Program and the Quality System Assessment Program, which were early attempts by the USDA to reduce fear of US beef in other countries. Those programs were not designed to include livestock markets, and the original USDA plans for the NAIS did not effectively incorporate livestock markets. Because some 40 million cattle are sold through these markets annually, and because they are typically focal points of communication for buyers, sellers, and brokers, this lack of information created a good deal of confusion surrounding the USDA's plans.

Still, there was no clear plan, and as Joe Roybal of *Beef Magazine* reported, one of the major concerns for the near future of the beef industry was the "USDA's seeming decline in credibility among various audiences and what it portends for the advancement of important industry issues and infrastructure such as the development of the National Animal Identification System" [17, p. 1]. Roybal's statement rang true because of the credibility that the USDA lends to any identification system, and because without USDA involvement, many overseas markets will always be unwilling to accept an industry-led plan.

Nevertheless, some matters were agreed upon by most industry organizations, including the LMA, NCBA, and the USDA. They agreed that any NAIS must

- Provide biosecurity for the national herd
- Identify vaccinated and/or tested livestock
- Identify animals in national and international commerce
- Protect animal data
- Provide added value
- Allow for efficient tracking of animals and disease outbreaks

1.1.6   USDA Strategies for Communication

When the real push came to educate producers, consumers, and members of the beef supply chain, the USDA's primary challenge was to communicate its plans and criteria for a national animal identification system to members of the beef community. Because the beef supply chain is so fragmented, this was not an easy task. Small producers often have limited contact with other producers, sporadic contact with livestock markets, and almost no contact whatsoever with large feedlots, stockers, or packers. This fragmentation was a problem for the USDA and other NAIS proponents because of the number of small producers that still comprise a major segment of the beef supply chain. Roughly one-third of the cattle sold annually in livestock markets come from small producers. Over 80% of those small producer operations have fewer than 50 cattle and approximately one-third of the national herd is in these small groups [7]. One-third of the national herd equates to approximately 29 million animals raised on approximately 915,000 farms and ranches. So, in effect, the USDA needed to communicate with nearly one million farmers and ranchers who have little or no contact with the majority of the beef supply chain.

Communicating with that number of small producers was a difficult task, to say the least. Small producers in rural areas might only communicate with other ranchers on an occasional basis and would most likely only sell animals at a livestock market a few times per year. Although the USDA did produce literature and brochures designed to explain the proposed NAIS and did attempt to distribute that information, many small producers and even livestock market owners did not have direct access to that information, and those that did tended to ignore it for various reasons (discussed later in this chapter). The USDA also sent representatives to regional meetings hosted by state organizations to promote NAIS, but those meetings were only attended by a small fraction of the 915,000 million small producers currently raising cattle in the United States.

Of course, industry alliances were (and are) able to disseminate information and opinions through word-of-mouth and through newsletter mailings to their membership very quickly. The LMA, NCBA, and R-Calf, just to name a few, are able to communicate with producers, buyers, and market owners through direct methods such as mailings and conferences not necessarily available to the USDA. This lack of direct communication was part of the reason that so many small producers and livestock market owners began to form their opinions concerning national animal policy in what might best be described as "coffee shop diplomacy sessions" on the basis of partial or biased information. Buyers, market owners, and sellers tend to communicate at local livestock auctions and at whatever meetings are available. But the informal nature of that communication led to the larger supply chain entities' representatives (who tend to travel more) passing on company or organizational interpretations of the NAIS to the smaller, more geographically limited producers and market owners.

As a result of these combined factors, confusion reigned supreme within the beef supply chain about what the proponents of NAIS were saying. That confusion prompted members of the beef supply chain to voice their concerns to their congressional representatives and to other members of the supply chain. In May 2006, a congressional subcommittee voted to withhold almost all funding for the NAIS until the USDA was able to better define the system's objectives. Then, the Congress approved only one-third of the USDA's requested funding for the fiscal year 2008 [18]. So what was it about the communications of NAIS supporters that did not resonate with livestock market owners, buyers, sellers, and producers? My research was designed to answer that question.

1.2 A New Approach to Studying Complex Communication Issues

I should point out that, as an employee of a technology firm, I had some interest in seeing the NAIS succeed. However, my primary objective during the time period covered by this research was to collect information about what members of the beef community thought of the proposed NAIS and how technology could best be adapted to their needs. The majority of information for this study was collected between 2005 and 2009.

In general, selecting research tools for an industry like the beef industry can be a daunting task. There are few research sources to consult from a communications perspective, and actual communication within the industry tends to be multifaceted. While some communications are paper-based, and others are posted on the Web, the majority communication within the industry is still of a much more personal nature, done either face-to-face or over the telephone.

In addition, the particular communications that I was interested in involved foreign subject matter, including new technologies and new methods of conducting business. Both were new to the industry and had not been studied previously. Therefore, I needed an approach to investigating this multifaceted communication that would analyze the effectiveness of various communications, while at the same time tracking how communications moved through the industry, affecting acceptance or rejection of technology as they evolved.

To address the issues at hand, I chose a model that would incorporate communication theory, linguistic principles, ethnographic principles, and theories of technological diffusion and communication networks. My intention was to include theories that would aid in the examination of the different methods of communication within the industry, while giving proper attention to the communication networks within the industry and maintaining a view of technological diffusion for its own sake. Certainly, technology diffusion has some life of its own, apart from communication, but in an environment like the beef industry, where technology (especially computer technology) does not yet have the stake that it has in other industries (even in 2014), diffusion and communication are almost inseparable due to the social nature of new technology diffusion.

In light of the beef industry's complex situation and the various methods of communication in play, four separate approaches seemed appropriate. Rogers and Kincaid's definition of communication networks [13] and Rogers' model of technological diffusion [19] were a good fit for the beef industry. Both are examinations of the way technological information is transferred and shaped. The two individual works go hand in hand in many ways, because they can be used to analyze the flow of information concerning the NAIS and technology within the beef industry and to analyze acceptance or rejection of that technology on the basis of principles of diffusion.

In addition, I needed an ethnographic methodology to understand the environmental and personal communicative factors that directly affect communications within the industry. Methods described by Fetterman [20] and Van Maanen [21] helped to establish an ethnographic perspective of my interaction with members of the beef community. Ethnographic methods were critical to the success of this analysis because of their ability to reveal some of the historical, demographic, and socioeconomic factors that affected the debate over technology and NAIS within the beef community.

Next, I needed to employ communication theory and linguistic principles to analyze documents from major proponents of the NAIS and technology, such as the USDA and technology development firms, and documents from those opposed to the NAIS and new methods of doing business. I needed to conduct a discourse analysis of message content and a general analysis of more informal communication. For these purposes, I chose to use Petty and Cacioppo's Elaboration Likelihood Model [22], and the linguistic concepts of implicature and pragmatics as defined by Grice [23], Blakemore [24], Schiffrin [25], Zwaan and Singer [26], and others.

Each of these theoretical constructs is distinct in its form and function, but together they offer a more complete picture of industry-wide communications than any of them could alone. Together, they form a tool that allows for analyzing communications that take place within a complex environment. I use the theories in the following ways:

1. Ethnography—to observe interpersonal communications and work roles within the industry
2. Communication Theory—to examine textual communications on the basis of their validity and persuasive appeal
3. Linguistic Principles—to examine textual communications for their attention to instances of implicature and their pragmatic consideration
4. Theories of Diffusion and Communication Networks—to provide an umbrella for the research as a whole and to frame research data

1.2.1 Ethnography and Diffusion in the Beef Supply Chain

The first phase of my analysis was to conduct ethnographic research within the beef supply chain so that I could analyze NAIS diffusion and communication within the beef industry. Livestock market owners were the primary focus of my interviews, because livestock sales are a primary point at which animals become "lost" (untraceable) and because these markets mark the point at which much of the resistance to technology is born. However, I also spoke with other members of the supply chain, including stockers and packers. Because of my professional experience with members of the beef industry, both as a casual participant and as an employee of a software firm, I had already developed contacts within the industry. These factors helped me gain direct access to my subjects and their businesses. In time, I was able to develop relationships with many of the livestock market owners and their customers and to engage in very frank discussions with them. I considered these personal relationships invaluable, both as a researcher and as an employee of a software firm, because of the informal communication networks that permeate the beef industry.

My initial conversations concerning technology were primarily with livestock market owners. The conversations were casual in nature and designed to gather general information about personal attitudes and beliefs about technology and animal identification. I spoke directly with 48 livestock market owners over the phone in 10 different states, including (in order of contact)

- Oklahoma
- Texas
- Arkansas
- Kansas
- Nebraska

- South Dakota
- North Dakota
- Minnesota
- Missouri
- California

My research was enhanced by a series of personal visits to livestock auctions. Between September 2005 and August 2006, I visited 35 of the 48 auctions that I had previously contacted by phone. All states listed above were included, with the exception of Minnesota and California. My personal visits to the livestock auctions allowed me to move beyond the owners' perspectives to include other livestock market staff, buyers, and producers.

As I visited with owners, staff, buyers, and producers, I aimed my questions at gaining relevant data in three main areas.

1. The level of willingness among participants to use new technology and animal identification
2. Their experience, if any, using online technology, RFID technology, various software applications, and databases
3. Their understanding of animal identification, new technologies, and where they gained that knowledge

Most of my conversations at the livestock auctions, aside from those with the owners, took place with the following types of individuals:

- Auction employees in charge of intaking animals from sellers
- Office managers of the livestock auctions
- Producers selling animals at the auctions
- Large-order buyers who regularly purchase many of the animals sold at the auctions

I chose these participants because, with the exception of the auctioneer, they represent each stage of the process of intaking and moving an animal from seller to buyer within the livestock auction and because they are the most likely users of any new technology that would accompany an animal identification system. I was interested in determining how these participants would differ in their experience with technology and in their willingness to accept a new ideology in the beef industry.

In addition to livestock market conversations, I spoke with procurement officers from major packing companies, order buying firm managers, USDA Animal and Plant Health Inspection Service staff, national and state legislators, chief executive officers (CEOs) of companies within the supply chain and parallel to the supply chain, and many cattle producers.

When considering innovations like those facing the beef industry, there is always more to the story than relating ideas and advantages to potential adopters of that technology. Technology diffusion is a social change that is directly affected by communication and one that should not be confused with the advantages of the technology itself. Hence, identifying the communication networks responsible for disseminating information about new technology and the NAIS was one of my top priorities in conducting interviews, both as an employee and as a researcher.

Because I was able to travel to so many livestock markets, I was able to speak with hundreds of individual beef industry insiders and to ask them about their knowledge of technology and NAIS. I could also speak with them about where their personal information was coming from and whose opinions mattered to them.

I asked the following questions to all formal interviewees. Other questions followed in most cases for clarification, but these questions were universal:

1. How would you rate your understanding of the National Animal Identification System (NAIS)?
2. What is your experience with technology including computer technology, ear tag scanning technology, and Internet/database technology?
3. How have traditional beef industry values and current business pressures shaped your opinions about both technology and the NAIS?
4. Do you believe that proposed new technologies will work within the beef industry?
5. Does the new business model of the NAIS fit the needs of the beef industry?
6. Where have you found most of your information about the NAIS and its accompanying technology?
7. How would the implementation of the NAIS and the accompanying technology affect your daily business activities and your personal responsibilities?
8. Who do you talk to about things like the NAIS, technology, and business concerns?
9. How has the information you have received concerning new technology and the NAIS affected your personal opinion of new technology and the NAIS?
10. Which information sources were most influential in shaping your current opinion of the NAIS?

My purpose in asking these specific questions was to gain an understanding of the culture within the beef industry in addition to specifics about the NAIS. As Fetterman writes, "Ethnography is the art and science of describing a group or culture. The description may be of a small tribal group in some exotic land or of a classroom in middle-class suburbia" [20, p. 1]. Van Maanen agrees, stating, "In the most general sense, ethnography refers to the study of the culture(s) a given group of people more or less share" [27, p. 4]. In this case, that description may be of a singular industry or segments of that industry, such as the beef industry. Fetterman points out that ethnography is, in many ways, like the task of an investigative reporter, except for the fact the investigative reporter seeks out the unusual, while the ethnographer seeks out the everyday, the usual. I sought to seek out the everyday person within the usual course of events in the beef industry.

As Van Maanen sees it, the insider's position in ethnography grants ethnographic research, "[…] a kind of documentary status on the basis that someone actually goes 'out there,' draws close to people and events, and then writes about what was learned" [27, p. 3].

Fetterman tells us that fieldwork is the most characteristic element of any ethnographic research design. Classic ethnography often requires extended periods of observation. However, as Fetterman also points out, in applied settings, long-term continuous fieldwork is neither possible nor desirable. The most important element of fieldwork is being there to observe, to ask questions, and to be in contact with your chosen culture. In the beef industry, that means spending time at livestock markets talking with market owners, buyers, sellers, producers, auctioneers, sale personnel, and other members of the beef industry.

1.2.2 Communication Theory, Linguistics, and Diffusion in the Beef Supply Chain

The second stage of my research was to analyze written communications concerning the NAIS and technology. My focus in analyzing written documents was upon the USDA and several NAIS proponents and detractors. Specifically, I was interested in the efforts to promote technology and animal identification within the beef industry, in analyzing the ways in which those ideologies were transmitted, and in comparing NAIS documentation with responses from industry alliances.

I chose seven primary texts for analysis on the basis of two considerations. First, I wanted the texts to be from sources directly and influentially involved in the NAIS technological debate. Second, I wanted the texts to be representative of the written materials that have been most widely circulated and read by industry insiders on the basis of my conversations with members of the beef community. The primary texts for analysis were these:

1. *Audit, Review, and Compliance (ARC) Branch Policies for USDA Process Verified Program 2004* [28].
   This document was designed to give readers an overview of general policies concerning the USDA Process Verified Program (PVP), which was the forerunner of the NAIS.

2. *ARC 1001 USDA Process Verified Program 2004* [29].
   This document is provided by the USDA Audit, Review and Compliance (ARC)branch as a guideline for applying for Process Verified Program status.

   The PVP was designed to assure customers and consumers that animal products have been raised and processed according to a prescribed set of standards designed to ensure safety and quality.

3. *The National Animal Identification System (NAIS). Why Animal Identification? Why Now? What First 2005* [30]?

This is a brochure designed by the USDA to persuade beef industry professionals to adopt the system. It was intended to offer insight into the merits and advantages of the National Animal Identification System.

4. *The United States Animal Identification Plan.*
   This is the initial report issued by the National Institute for Animal Agriculture (NIAA), which was the nonprofit organization that organized a task force of USDA and industry professionals in 2003 to draft a plan for the NAIS. The draft was issued in January, 2004 [31].

5. *Livestock Marketing Association Comments on the US Animal Identification Plan 2005* [32].
   The Livestock Marketing Association's response to the NIAA's report.

6. *National Cattlemen's Beef Association NAIS Industry Proposal White Paper.*
   The National Cattlemen's Beef Association Response to the NIAA report, issued in February, 2005 [33].

7. *R-Calf USA 2006 Position Paper: National Animal Identification System* [34].
   R-Calf's response to the National Animal Identification System.

Together, these documents provide an overview of the USDA's pre-NAIS responses to international pressures concerning animal identification, the USDA's plans for the NAIS, and the responses to that plan from the largest industry alliances. Using theories of communication and linguistics to analyze these documents was critical to identifying misunderstandings as opposed to simple disagreements. My goal was to use communication theories to discover what strategies were used to attract and persuade members of the beef community, which strategies were or were not effective, and what strategies were missing.

My hypothesis was that gaps in language commonality and poor use of contextually based persuasion techniques may have led to some of the miscommunications that occurred during the initial debate over the NAIS and technology within the beef industry. Therefore, I began with Elaboration Likelihood Model (ELM) principles by investigating whether

1. an individual message was likely to be considered using critical thinking skills;
2. readers would be motivated to consider information presented by the source of the message;
3. typical readers would be able to understand the information presented by the message;
4. positive or negative cues were likely to be adopted by the reader in the absence of central route processing;
5. bias may have affected persuasion.

The ELM, developed by Petty and Cacioppo [22], provides a unique perspective on relevance and the likelihood of cognitive consideration from members of communication networks. According to the ELM, the question is this: What is it about the content of a message that gives that message the ability to persuade listeners to really listen, so that actual persuasion can take place? The ELM was originally developed specifically for examining communication intended to be persuasive, and it can be applied to both oral and written communication as a measure of both persuasiveness and effectiveness.

The model is built upon several basic assumptions about communication. One is that users want to form correct attitudes. As the authors point out, we are motivated to hold correct opinions because of their influence on our mental and physical well-being [22]. While the correctness of an opinion is inherently subjective, our opinions and attitudes can always be judged against other standards that allow us to evaluate our attitudes and behavior.

Another basic assumption of the model is that, although we want to form and hold correct opinions and behaviors, the ability to do so varies by person and by situation. In other words, the amount of effort we are willing to expend toward forming correct attitudes varies on the basis of a person's ability to consider the facts of a message, the context in which the message is received, and the relevance of the subject matter for the message receiver.

Finally, the model points out that we are simply unable to think critically about every persuasive communication we receive. Consider for a moment the number of advertisements, political messages, written opinion pieces, newspaper articles, and television reports we see in an average day. It would be impossible to consider all of those messages critically and get anything else done. We would spend all of our time in a futile effort to form correct opinions on an endless variety of persuasive subject matter.

On the basis of those postulates (and several others), Petty and Cacioppo outlined two routes through which persuasive communications are channeled. The central route is the route that is used when a message recipient is both motivated and able to think carefully and critically about the information being presented. The authors define this route as the route used by message recipients when they are both willing and able to elaborate (think critically) upon the information presented. As they put it, "By elaboration, we mean the extent to which a person thinks about issue-relevant information" [22, p. 7]. Elaboration likelihood is high when the conditions surrounding the message are not distracting, the motivation of the recipient to correctly process the message is high, and the individual abilities of the message recipient are conducive to elaboration of the content matter of that message.

If the message recipient is unable or unwilling to use the central processing route, the recipient will be more likely to use the peripheral processing route. In this case, the recipient is more inclined to use positive or negative persuasive cues in context to determine the validity of the message. In this case, the elaboration likelihood is said to be low. In other words, the message recipient either does not possess the ability to process the information presented or is not motivated to do so and will turn to peripheral cues to form an opinion.

The central processing route, which requires critical thought, will produce more affective and lasting changes in attitudes and behavior, while attitudes and behaviors formed through the peripheral route tend to be less enduring and affective. As to persuasion, we can assume that users are persuaded to the extent to which they find communication usable, understandable, and relevant to their personal context and goals. It makes sense, then, that if these conditions are met, users will be more persuaded of the value of new technology and business methods.

1.2.3 Linguistic Textual Analysis

Next, I turned to linguistic theory to analyze the same documents on the basis of implicature and pragmatics. My goal in doing so was twofold. I was looking for instances of both implicature and pragmatic assumption that would hinder communication between the USDA, industry alliances, and the average beef industry professional. Specifically, I was first looking for the number of implicatures required of the reader, defined as the number of times the reader is required to fill in information in accordance with the Gricean maxims. Second, I was looking for pragmatic acknowledgment, defined as the amount of context given to the reader about situational aspects of the text (i.e., in what situation you would be expected to perform certain operations or understand certain ideological concepts). I was interested in what the texts assumed the user's context to be, the number of implicatures required by the texts, and their use of lexical items as either abstract or definite concepts.

Implicature, as described by Grice [23], is that part of the meaning of a sentence or text that is left to be filled in by the reader. Implicature is a function of what Grice calls the cooperative principle. The cooperative principle is based upon general assumptions about conversations stated as maxims:

1. Quantity—communication should be as informative as necessary, but not more so.
2. Quality—communication should not be deliberately false or one for which you lack sufficient evidence.
3. Relation—communication should be relevant to the subject at hand.
4. Manner—communication should be clear, orderly, unambiguous, and concise.

These maxims are what we expect in conversation, but as Grice points out, we often deviate from these maxims in any number of ways. For example, we rely upon violations, or *conversational implicatures*, on the basis of our context, to make conversation easier. In other words, our conversational situation often creates a need to violate the maxims.

For example, look at the following hypothetical conversation.

Conversation 1

Statement: "Let's go out to eat tonight."
Response: "We don't have any money."

From this conversation, we are able to deduce relatively quickly that the couple cannot *afford* to go out to eat, even though the response might have an entirely different meaning without the initial statement. In Gricean terms, this conversation represents a successful implicature based upon the maxims of conversation, which in essence state that participants in a conversation operate cooperatively under the assumption of a set of rules. The two speakers are able to create this successful implicature because of their shared context, goals, and background.

This cooperative principle helps us to conduct communicative interactions within spoken and written contexts and operates under the assumption that both parties have similar goals for the communication, that the amount of information is enough, but not too much, and that the information should be both factual and relevant to the conversation. In the conversation above, the initial speaker is able to make sense of the response because of shared background and contextual knowledge with the respondent. We understand that going out to eat requires money. Thus, even though on the surface the response seems to be irrelevant, we understand its implications.

It is in this respect that implicature becomes necessary in communication. For example, let us revisit the conversation above. Without the implicature it might look like Conversation 2.

Conversation 2

"Let's go out to eat tonight."
"We don't have any money."
"So what?"
"So we can't go out to eat."
"Why?"
And so on…

Thus, knowledge that having no money means that the couple is unable to eat out is critical to the conversation. Without it we might be unable to complete this explanation before dinner! Recent linguistic theory assumes that implicatures are processed simultaneously with text, or sound in the case of spoken conversation, making the reader instantly aware of complex nuances that change the meaning of the communication [35]. This is the reason we understand the implicature required by Conversation 1 and is the reason we rely on implicature so completely in written texts. Within the beef industry, we can begin to see why this might have an impact on communications about technology and the NAIS. For example, what if Conversation 2 looked like this?

Conversation 2A

"Let's track our animals after they leave here."
"We can't."
"Why not?"
"We would need a database."
"What's a database?"
And so on…

However, there is one major difference between the implicatures required in the example above and those required in many written documents. The implicatures required of readers of written documents are often not based upon the maxim of quantity, which states that the reader's background knowledge must be taken into account so as not to give too much information. Violation of this maxim is often necessitated by single words that function as lexical items for other concepts or terms. Lexical items can be briefly described as the mental image or abstract thoughts produced by words, and they are closely tied to context. For example, when we read the word *paste* in a situation involving computer software, we do not think of glue. More likely, we think of the concepts involved with cutting and pasting procedures in word processing applications. Thus, a whole range of concepts and activities that may be required of the reader are encapsulated by a single term. To understand that term, we must understand not only the definition of it in this context but the actions that pertain to it in the current context.

Pragmatics, which, broadly defined, equates to the study of language in context, is concerned with the difference between what a sentence of text says and what the reader takes as its meaning. The perceived meaning is usually based upon the situational context of the communication. Thus, pragmatic considerations play an important role in shaping our interpretation of written and spoken language, which brings us to the critical role pragmatics plays in communication. While *pragmatics* can be defined as the study of language use in context, *context* can be best described as the situation in which the reader or listener will be using the information provided plus the relevant background knowledge the user brings to the situation. This definition is perhaps narrower than most definitions of context within linguistics, but more closely aligned with the concept of context within technical communication. Let's return to Conversation 2.

Conversation 2B

"Let's go out to eat tonight."
"We don't have any money."
"So what?"
"So we can't go out to eat."
"Why?"

If we know that both parties are solvent, it might look different.

Conversation 3

"Let's go out to eat tonight."
"We don't have any money."
"So we'll stop at the ATM."

Once again, the conversation makes sense, as long as we are aware that the statement "We don't have any money" only means that the couple does not have any money *with them*. This meaning can be inferred because of our background knowledge concerning their wealth, which is part of the context in which we use the information provided.

Pragmatics also considers the situation in which a conversation takes place. For example, if Conversation 3 were held in a raft adrift on the ocean it would be humorous, even though the couple might be starving. Why? Because the context in which the conversation was held would signify that the remark was meant to be humorous. Therefore, we must consider both context and background in communication analysis.

In the beef industry, it is easy to see that we might find problems in communication on the basis of both implicature and pragmatics. Terms like *database* and *software*, which are not common to the beef industry, might create a need for implicatures that cannot be created by readers. Furthermore, even if the reader has some familiarity with the terms, the fact that they have not traditionally been used in the context of the beef industry may create problems with pragmatic consideration of their meaning.

I determined the number of implicatures present in each of the seven publications selected for analysis by using the following criteria and by assuming the abilities of an average market owner or producer on the basis of my ethnographic investigation:

1. How many times is the reader required to fill in information as a result of implicature that would be beyond the ability of the average beef industry reader (as determined by my conversations with them)?
2. Does the text give indications as to the expected background knowledge of the reader?
3. Does the text offer specific contexts to use in interpreting the text?

For example, the following passage from the USDA NAIS brochure creates implicatures on the basis of a lexical item and on the basis of assumptions made about the background of the reader.

Example 1

Database systems must be developed and maintained, equipment must be purchased, animals must be identified and tracked, programs must be monitored, and labor is needed for all of these activities [30, p. 1].

The language used in Example 1 requires the reader to identify the term "database" through implicature or background knowledge. This term is not explained within this brochure, at least not to the extent needed by a novice technology user. In effect, this segment rests upon assumptions about user background knowledge concerning database technology that are probably unfounded. This assumption means that the reader will either have to stop at this point to find what the term means or continue without the knowledge that the correct implicature would provide.

I was also looking for provision of user context and goals within this type of passage. For example, does the text tell us *why* we might want to use a database? Or where that database might be used? In this case it does not. Example 2 offers another implicature from the USDA PVP policies. In speaking to what is required of PVP-endorsed companies, the USDA observes as follows.

Example 2

All evaluations and reevaluations of must be in accordance with the *Principles of Auditing* as defined in ISO 19011:2002 guidelines for quality and/or environmental management systems auditing [36, p. 2].

Again, Example 2 requires the reader to possess knowledge about several terms, including "Principles of Auditing" and "ISO." Some readers may understand that auditing is related to accounting and accountability and that ISO refers to the International Organization for Standardization. However, on the basis of my interviews, most beef industry professionals (and most people) are unfamiliar with these terms and certainly have no idea what the International Organization for Standardization does or why it is important. When we look at the actual implicature drawn by this reference, we see that the writing assumes knowledge sets that are probably unrealistic.

1.2.4 Diffusing Innovations in the Real World

Everett Rogers' [19] book concerning diffusion of innovations provides a valuable tool for analyzing traditionally nontechnological industries like the beef industry, where proponents of the NAIS are seeking to diffuse new information technology and innovations. Rogers also worked with Kincaid [13] in research about communication networks; because communication within the beef industry is highly networked and affected by multiple perspectives, this second text provides an additional tool for analyzing communication networks. Together, the two texts form a powerful tool capable of analyzing both technological aspects of diffusion and the human side of communication that can dominate technological discussions.

Technology, in Rogers' description, is a means of reducing uncertainty about cause and effect. The potential benefits of the innovation motivate us to learn about the innovation, while the possible consequences of adoption limit our acceptance. In the case of the beef industry, the proposed innovation is the NAIS and, consequently, a new method of doing business. The benefits of that innovation could lead to a safer food supply, better demand for products, new methods for tracking animals, and host of other potential benefits. But, the unknown facets of the plan, combined with known disadvantages, are a constant pull away from the innovation's adoption. In this case, the confusion is compounded by the fact that the new technology also brings a new business model with it.

The balance of these two considerations determines the willingness to accept new technology. In this sense, innovation acceptance or rejection is essentially a process of information seeking and experimentation. When confronted with innovations, people typically begin by asking a series of basic questions designed to help accept or reject technology:

1. What is it? (e.g., What is the NAIS?)
2. How does it work?
3. Will it change the way I work?

4. Why does it work? Or why does it fail?

5. What are the possible negative consequences of its use?

6. What are the possible advantages of its use?

To answer these questions, potential adopters go through the process of diffusion, which Rogers presents as a combination of four elements:

1. Innovation—during which new technology is compared to existing technology. Questions that users ask during this phase include these:

   (a) To what degree is it better than what we have now? (relative advantage)

   (b) Is it consistent with the values, experiences, and needs of our current situation? How much do I have to change?

   (c) How difficult is it to understand and to use the innovation?

   (d) To what extent can it be tried and evaluated?

   (e) Can we know what the current results of its use are? And are we able to observe trials?

2. Communication channels—the means by which messages about the innovation get from one person to another. But the messages themselves are not the pure information we would often like them to be, because most people within social networks rely upon the subjective evaluations of others to form opinions about innovations.

3. Time—which can be measured through the completion of five stages:

   (a) Awareness of an innovation and some knowledge of how it functions

   (b) Formation of an opinion about the innovation

   (c) Active acceptance or rejection of the innovation

   (d) Implementation of the innovation (if put to use)

   (e) Confirmation or reversal of a decision because of experience or new information

4. The social network—used to form opinions of new innovations. Most individuals seek the opinions of others whom they respect or trust or both. In the case of the beef industry, those "others" typically are peers in the industry and representatives of industry alliances.

1.2.5 Diffusion and Communication Networks

There is an old joke that we only read the user's manual as a last resort. Although formal structures do exist for communication flow in most industries, there is an informal element to those communication networks that often defines much of the communication that takes place. In other words, we ask a friend first. In the case of the beef industry, and others I suspect, those informal networks are actually separate networks that operate independently of the formal networks that are organized around the supply chain. Social networks form within industries and control the flow of information apart from formal structure. Eventually, certain patterns form. In a sense, they resemble the path water takes in a stream as the flow becomes normalized over time. A channel forms, and almost all water follows that path to its ultimate source. The same can be said of information within the beef industry. As individuals become comfortable with other individuals as sources of information, information tends to flow from the same source repeatedly. At the same time, this well-developed channel begins to exclude other sources of information.

Rogers' and Kincaid's work on communication networks is an attempt to explain communication in real-world contexts. In their view, communication in social networks often involves two or more persons sharing information simultaneously in order to reach mutual understanding. They call this process convergence. This model places emphasis on the relationships between those exchanging information rather than on individuals themselves as units of analysis. Members of the beef community receive information about new technology, but their communications are anything but linear, and those communications are directly affected by personal and social variables. Therefore, the beef industry more closely resembles a network model of communication than linear models.

In the network model, information is not a commodity to be transferred from one individual to another as one would hand someone a glass of water, but an interpersonal act complicated by special relationships, psychological bias, and mutual causation. Communication, then, is a process that is affected by the individuals involved, their experiences, their predispositions, and their personal histories. Certain individuals within the network routinely communicate with other individuals within the network, and patterned flows of information are created. The most important step in determining how a communication network affects behaviors on the basis of information is determining which individuals (or organizations) are affecting others within the system.

Although a message may flow from one source to an end point, that information will be interpreted and often reshaped as it travels through established paths of communication. We might think of these interpretation points as ports along the stream metaphor. In the case of the beef industry the LMA, NCBA, R-Calf, and other similar alliances operate as these interpretive ports, as do influential businesses and individuals.

1.3 Results of My Investigation

Most readers have probably never been to cattle auctions, which are typically held at livestock markets. But if you want to get a feel for what is really happening in the beef industry, there is no better place to start. Livestock markets vary widely. Some are held in their original buildings (which may have been built decades ago). Others are newer, with modern office space separate from the auction building. Some are in urban settings like Oklahoma City, while others are literally hundreds of miles from anything resembling urbanization.

Still, they share some characteristics regardless of their location. They are dusty, noisy, active places. In many parts of the country they serve as a focal point of both agricultural business (which is the only business in some areas) and social interaction.

Most auctions have a café that also serves as a coffee shop. On sale days, which are typically weekly events, agricultural professionals from the area gather at the livestock market to discuss current events, buy and sell animals, and to keep abreast of what is happening in different agricultural markets. It is both a social and professional endeavor. News and ideas are exchanged, business partnerships are formed, and services are procured. There is a generally lighthearted atmosphere outside of the auction arena, but once the sale begins, the market personnel become more serious in their efforts to manage a fast-paced environment where big money changes hands.

My conversations with and observations of livestock market staff revealed a great deal concerning both technology and the NAIS within the beef industry. While the results were predictable to someone who has been involved in the process, they may offer some surprises to those with little experience in dealing with inexperienced technology users. My initial results from these conversations and observations with market owners, staff members, producers, and buyers can be summarized by category.

1.3.1   Alice at the Auction

Cattle auctions are busy, fast-paced environments, with numerous employees performing a variety of tasks and both buyers and sellers trying to keep up with the action. At the head of the auction business apparatus are the office managers. Auction office managers, on the whole, have had more direct experience with technology than any other personnel. They are typically called upon to perform several professional jobs at once including accountant, human resource manager, administrative assistant, communications director, and a general point of contact for anyone and everyone.

The first office manager I met was named Alice, and after a while I began to refer to all of them as "Alice." They are strikingly similar as a group, being largely overworked, underpaid, always female, and probably the most knowledgeable people at the auctions in terms of daily happenings. They tend to be generally cheerful in spite of their workload, but one quickly finds that in addition to being intelligent and cheerful they are "not to be messed with." Most of them have a reputation for their ability to deal with everything from angry customers to delinquent accounts.

Office managers were usually more willing to discuss technology than other auction personnel. They were so much more willing, in fact, that my habit became to look for "Alice" upon my arrival at every auction. On the whole, the office managers I spoke with were by far the most accepting of the idea of new technology and business methods. They seemed to be the most willing of the participants to explore the potential benefits of the NAIS and the accompanying technology. My conversations with them revealed that most have had experience with a string of data collection systems over time. However, most of these data collection systems were nothing like the advanced technology that would be required by the NAIS. In truth, the data collection systems used by most auctions I visited in 2006 were either severely outdated (many still based on a DOS platform) or were built to work on outdated networking systems. Of the 48 auctions I visited, only 15 were in the process of updating their systems, or they were already running current Microsoft® operating systems.

The many tasks that office managers perform during animal sales are often hampered by the general technological inefficiency that accompanies them. For example, my observations showed that livestock market staff members typically enter data manually for each animal as it is sold. In many cases, information is written by hand on note-size forms as each animal or group of animals is sold. The forms are then taken in batches to the auction office, where they are entered, by hand, by Alice and any help that she may have into whatever accounting/spreadsheet system exists. Later, they fax these data to the buyers, which is an inefficient and time-consuming process. Many office managers would be happy to automate these processes if technology could help. One office manager told me that she would be able to die happy if she could just eliminate entering sale data manually. However, even the office managers, as a group, are far removed from the concepts of networked machines, databases, servers, and Internet data transfers.

In effect, the office managers run the auctions on sale days. The owners are rarely involved in the accounting and data transfer that accompanies a fast-moving auction. Instead, they are busy dealing with customer concerns. But, most office managers are also responsible for training their sale-day office staff, and my conversations with them revealed a lack of confidence among office managers concerning training themselves and others in the use of new technology. Most would be unable to direct such training and would not have time to do so anyway. Also, they worry that if this training were to go poorly after hardware had been installed and new methods of business had been implemented, the auction could be severely hampered and they would be blamed.

Finally, it was clear that the information that the office managers did have about the NAIS and the technology that would accompany it had been gained from largely informal interactions with their respective market owners, animal buyers, and animal sellers. Their opinions of the NAIS tend to reflect those of the other main players at the auction, and their opinions vary. People in some areas tended to have more optimistic feelings about the NAIS than those in other areas, but more specifically, producers and market owners had different opinions from auction to auction. While those differences during my investigation may have been only varying levels of confusion and distrust, the differences show in the opinions of the office managers.

1.3.2 Backstage at the Sale Barn

Of course, before animals could be sold at an auction in accordance with the NAIS, and long before the office manager would be involved, they would need to be fitted with an electronic ear tag or some other device containing unique identifying data. Tagging done at the sale barn would need to be done in the pens behind the sale barn in a specialized chute designed to keep animals stationary for their protection. Normally, these chutes are used to immobilize animals for injections and other veterinary procedures, but they would also be used to immobilize animals so that RFID tags could be placed in their ears. Those responsible for both attaching the tags and reading them would likely be the livestock intake personnel.

Not surprisingly, the intake personnel at the auctions who would be responsible for collecting identification information have, for the most part, extensive cattle knowledge

FIGURE 1.2. The back of a cattle auction is where cattle are sorted and kept before the sale.

and almost zero exposure to the type of technology required by the NAIS, or even personal computers in many cases. My conversations with intake workers revealed that they had no real fear of technology, but they did not trust it. Most received limited, if any, help with the technology at their markets and would prefer direct instruction with new technology, as they have no patience for manuals.

The "back of the barn," where the intake workers spend most of their time, is a series of metal frameworks and chutes that are used to control animal movement, group animals together, and coordinate arrivals. It is a dirty, noisy, chaotic environment with hundreds or even thousands of animals on hand on any given sale day. Figure 1.2 shows the back of the sale barn in Kilkenny, Ireland. The area itself is usually covered, but rarely enclosed, and most of them have walkways over the many animal pens to speed movement from one side to another. Some employees are usually on horseback herding animals, while others tend the gates that control animal movement.

It is not surprising, then, that the idea of a laptop computer or similar technology in this area of the auction often makes intake workers laugh. The very notion of advanced technology in such a traditional, hands-on environment seems almost ridiculous, even to an outside observer. In addition, intake workers were extremely concerned with the additional animal processing time that NAIS technology might require during an auction. Auctions often have hundreds, if not thousands, of animals to sell on a sale day, which means that speed is critical to their tasks. Intake personnel must move quickly to keep up with the demands of the auctions. An addition of even 10 seconds per animal, for example, could extend an already lengthy sale by hours, causing added stress for all involved.

However, it should be noted that many animals arrive one or two days before the auction, which would allow for these tasks to be done over time with those animals. Still, the addition of new technology, and therefore new tasks to the sale, would require extra staff and new skills during that time. And, in their defense, the USDA has initiated a large effort to identify private ranches with premises identification numbers so that animals could be tagged before even coming to the auction.

But, even if the animals were tagged at the ranch before arriving at the market, the tags would need to be read with specialized wands or other equipment to collect the data stored on the tag. Therefore, the employees working in the back of the livestock auction would be responsible for first RFID tagging the animal's ear (if not already done) and then reading that tag into the data collection system (probably on a laptop computer or specialized data collection device), at a minimum.

Nonetheless, as with the office managers, some intake workers favor a system that could automate tasks and save time. I say some, because most intake personnel are perfectly happy with the way things are. The system in place has worked for them for the past 50 years, they say, so why fix what isn't broken?

1.3.3 Buying the NAIS

Of course, after an animal has been processed in the back of the barn, it must be sold. Without a sale, there is no animal movement, which is what the NAIS was designed to track in the first place. That is where the order buyers come into the picture. Order buyers typically move from auction to auction on different days of the week, purchasing large numbers of cattle that are then shipped to other locations. Sometimes those locations are feed lots, where animals are fattened for harvesting. Sometimes those locations are other ranches, where the animals feed, live, and produce young ones. And sometimes those locations are packing plants, where the animals are harvested and converted into the packages we find at the grocery store.

Although order buyers are certainly not the only buyers at the auctions, their firms operate by taking orders for cattle from many places and then sending buyers to auctions to fill those orders, which often range in the hundreds of thousands of head per year. Additionally, packers often send their own buyers to auctions to fill quotas for production. Most of a typical order buyer's clients are either packers, stockers who intend to let the calves graze on grass or wheat as they grow, or feed yards, which hold thousands of cattle at a time while they are fed on grain.

Because the buyer and/or the staff of the buying company are responsible for quickly and accurately transferring data that usually come by fax into a spreadsheet form of some type (which has to be done manually), most can see clear benefits to new technology. However, I spoke with several large order buyers who have experimented with some type of electronic data transfer and have experienced setbacks because of their lack of experience and lack of information.

One story I heard concerned these initial attempts to streamline data transfer. The company in question was using an automated data transfer to dump purchase information directly into a Microsoft Excel spreadsheet. At the same time, they were experimenting with buying tagged, identifiable animals. While there were several distinct sets of data being collected in the spreadsheet, one of those columns was reserved for the unique animal identification number from the tags.

The experiment went well at first, until the company's office became convinced that all of the unique identification numbers transferred with the animals were identical. This confusion resulted from their staff failing to realize that after exporting the data to Microsoft Excel, the numbers (being very long) would be automatically expressed in scientific notation by Excel. The results caused a great deal of confusion and were typical of the many small problems that I heard of while discussing technology and the NAIS with order buyers.

In general, the buyers themselves are more concerned with getting quality animals at a good price than the technology that might accompany them. While their home offices may benefit from technology, provided that they understand how to use it, most buyers I spoke with saw the NAIS as fraught with potential problems and hassles.

Furthermore, what understanding they did have of the NAIS had been gained from sources other than the USDA. Unlike office managers and intake workers, buyers tend to have more contact with other elements of the supply chain. They are in between the producers, office managers, intake workers, market owners, and the rest of the supply chain. In a sense, they operate between small business and big business, making them excellent sources of information and perfect change agents, or so it would seem.

Strangely, they are not. The majority of buyers I spoke with were just as reluctant as anyone else to change their way of doing business, especially because they tended to have even less understanding of how the NAIS would affect their personal responsibilities and daily lives. In the end, buyers were less likely to have an opinion (or at least to express one) than any other group I spoke with. They simply were not interested.

That sentiment makes them more like the sellers of those animals than different. The sellers at the auction are typically the producers, or those who have raised the animals on their land. Sellers want to earn top dollar for their animals, while buyers want to pay as little as possible. But the two groups share a common communication network (at least in part) and both see the NAIS as a potential problem in terms of both daily life and profits.

1.3.4 Down on the Farm

Undoubtedly, the largest group to be affected by the NAIS would have been the producers—those persons involved in raising animals for sale at markets. My initial conversations with producers were unenlightening. Those I initially contacted saw any form of a National Animal Identification System as an intrusion by the government into their private affairs and nothing more. Like the animal intake workers, the producers I observed and spoke with had almost no ties to technology. Most producers I spoke with would have no use for any type of NAIS system unless it had proven economic benefits or was mandated by the federal government. They are, for the most part, people who have been raising cattle for most of their lives in exactly the same way.

Generally speaking, they are not interested in changing now unless they are forced to change or shown significant profits. In fact, the only example of sustained producer interest that I witnessed during my observations was at an auction in Texas, where a major order buyer began offering premiums for source and age-verified calves (which is proof of age and where the animal originated only). When the premiums were consistent, producers consistently provided the information that was needed to source and age-verify their calves. But when the premiums stopped, the producers selling animals at the auction became unwilling to provide animal data or to go out of their way to comply with NAIS goals because they had more profitable things to do with their time. Several times, I observed producers asking about premiums, only to change their minds about participating when an auction worker attempted to explain the process to them. One producer in particular told me,

> "Raising cattle is a 24-7 job already. If I can't make money with this thing then I can't use it."

He meant, of course, that if there were no profits to be had from animal identification, he had better things to do with his time.

His attitude was typical for producers. Although in my experience they tend to have the least understanding of the system and technology that would be required by the NAIS, they would potentially be the most affected group by its implementation. Because of this contradiction, producers who had been exposed to NAIS information tended to be interested in what it would mean for them. However, their information was typically gathered in places like the coffee shop and from sources that included market owners, other producers, and industry alliances like the NCBA and R-Calf.

Many producers are not "in the industry" as a lone profession. Small producers often have other full-time jobs that provide stable income, health insurance, and other benefits. That does not mean, however, that they are not dependent on the income they receive from their livestock operations. It means that they are unable to devote large amounts of time to understanding the NAIS and new technology. The result is that they turn to quicker, trusted sources of information to form their opinions. Market owners (those who own the auctions where buyers purchase animals) are the people in the area who devote time to interpret information about industry developments, so producers often turn to them for information. The fact that many of them do not turn to the USDA for information is interesting, but not surprising, given the principles of communication we have discussed to this point. In truth, producers turn to whatever sources of information are available.

The accuracy of those sources is questionable in many cases. I overheard countless highly opinionated conversations on the subject of the NAIS between producers, buyers, market staff, and office managers. Many of those conversations were based, at least in part, on either faulty or partial information. It was evident that during those conversations the participants were not dealing with information as presented by the USDA. The information that drove those conversations had often been interpreted, repackaged, and passed from one communicator to the next or one communication network to the next. Stops along the metaphorical stream had already altered the original information, because the original information had not been understood, because it was incomplete, or because of biases inherent to information gatekeepers within the industry's communication network. As shown by ELM, when presented with incomplete or confusing information, we tend to rely on peripheral cues that are more readily available. That tendency was a major component to the misunderstandings and resulting backlash that plagued the NAIS among members of small communication networks. Nevertheless, opinions were formed, and they were not unlike those formed by beef industry insiders with access to more information.

1.3.5   Interviews with Members of the Beef Industry

Although my casual conversations and observations with members of the auction community provided valuable insight into general feelings about the NAIS and technology, they were limited in their ability to provide detailed information. I needed to conduct more formal interviews, starting with livestock auction owners. Predictably, my interviews with livestock market owners echoed my original conversations with and observations of them in that they felt generally uncomfortable with the NAIS and new technology. However, they did provide information about the communication barriers and communication networks that have affected the debate surrounding the NAIS and the interpersonal and written discourse that have shaped opinions of the NAIS within the beef community.

In addition, my interviews with other members of the beef supply chain provided an external perspective that also proved insightful. My interviews made several things clear. First, the beef community as a whole still had no idea what to expect from the USDA concerning the NAIS. Second, livestock market owners did not fully understand either the ideology or the technology that would accompany the NAIS. Finally, communication networks were influencing members of the beef industry to a much greater extent than the official communications from the USDA and other proponents of a national animal identification system.

In spite of those general attitudes, there were more technologically progressive elements of the industry, even if they did not support the NAIS plan. One of the most illuminating interviews I conducted outside of the livestock markets was with a senior executive for Cargill Meat Solutions, which is one of, if not, the largest beef-packing companies in the world. Compared to producers and livestock markets, corporate packing companies are much more technologically sophisticated. Many of the large packers trade cattle like any other commodity, and to visit their offices feels much more like a trip to any other large corporate entity than a trip to a ranch or a livestock market. He spoke with me at length about the problems facing the NAIS.

The executive's attitude could best be summarized as cautiously optimistic although disdainful of the USDA plan. He reiterated to me that the beef industry still had no clue about how two of the domestic mad cow disease cases actually started and that this was a definite problem. However, he was adamant that the plan for the NAIS offered little profit return on investments that would have to be made to implement the NAIS plan.

He believed that the USDA would never be able to justify a national system solely on the basis of risk and that they must provide incentives other than a "feel good attitude" for participants to justify a national system. Otherwise, he believed that the only way to gain widespread acceptance would be to mandate participation, which would require political action. In his opinion, political action would be hard to gain because of the

USDA's long history of ignoring the business needs of the beef industry and industry alliance ties to state and national legislators. He said that although some companies like McDonald's have actively pushed the NAIS and have even offered to pay premiums for source and age-verified animals, they were alone for the most part among large retailers, because others were unwilling to commit to a plan backed by the USDA that is so ill-defined. Finally, he claimed that the plan changed from month to month, continually being swayed by different special interests, which led to a continually changing story from the USDA about what to expect of participants.

My conversations with other food company executives were similar. I had numerous conversations with personnel from other companies. Some of those included Smithfield Foods®, Tyson Foods®, several order buying firm managers, order buyers, USDA Animal and Plant Health Inspection Service staff, national and state legislators, CEOs of companies within the supply chain and parallel to the supply chain, and many cattle producers. Several themes emerged which were similar to those found in my interviews with livestock market owners: namely, that the USDA's plan had been ill-defined from the start, that it had been communicated poorly, that there was no general consensus as to what the plan would be or if it *ever would* be, and that the information that had been provided was confusing.

1.3.6 Interviews with Livestock Market Owners

Although I spoke with many different members of the beef supply chain, most of my personal discussions and interviews were conducted with livestock market owners. My initial conversations showed that livestock auction owners already knew more than they would like about the NAIS. Although they were interested in the potential benefits of the system, they were concerned with the economic implications of any NAIS plan. In addition, aside from this general attitude, most owners expressed no real desire to become familiar with operating NAIS technology and no desire to become technologically proficient in the details of the system. Most of the owners I spoke with in 2006 indicated that they did not use computers on a regular basis and did not want to. Most of those owners still own the same auctions today.

I did not get the impression that they were completely against technology. Indeed, owners described technology such as automatic scales and veterinary technology as indispensable to the cattle industry. Yet, as a group, they showed no ability to be involved in operating the technology that would accompany the NAIS—especially computers. In fact, 80% of the owners I originally contacted by telephone told me that they did not understand NAIS technology and did not feel they had been given much information about what would be required to implement the system. One owner told me that a certain number of market owners would need to die before advanced computer technology would ever become commonplace in markets! In general, they believe that any NAIS plan would result in more work for everyone and additional expenses.

Moreover, it was clear from my initial conversations with owners that they did not understand what was being proposed and were extremely reluctant to commit themselves to any ideology or course of action until they knew more. Furthermore, they were reluctant to commit to any course of action until market owners had established a consensus and until the physical and technological parameters of any such plan had been firmly established. They were also genuinely concerned with the challenges that new technology and the NAIS would pose for their staff members.

Of the 48 owners I had originally spoken with and/or visited, many were also willing to participate in interviews concerning the NAIS and technology. I knew some of those who refused to participate well enough to ask, "Why?" The most common answer was that the owner believed that he or she would have nothing to say on the subject. Further investigation revealed that many of those who did not participate felt that they had not spent enough time personally investigating the facts to offer insightful comments. I attributed their reluctance to apathy at first, but as my study progressed, I began to see their reluctance as confusion, more than apathy. My interviews provided direct insight into the reasons behind this surface-level apathy. Again, here are the primary questions asked of market owners and summaries of their responses to those questions. The responses on the whole were surprisingly similar:

1. *How would you rate your understanding of the National Animal Identification System (NAIS)?*
   All owners I spoke with had some understanding of the plan, yet very few claimed to have a solid understanding of the mechanics of the plan. They were also quick to add that the USDA did not know what it wanted to do, either. Naturally, I spoke with some owners who had spent more time investigating proposed regulations than others, but the general consensus was that owners did not believe that anyone understood the plan, most of all themselves. They were also frustrated by the lack of details available from "official" sources and the confusion surrounding the technology of such a far-reaching plan.

2. *What is your experience with technology including computer technology, ear tag scanning technology, and Internet/database technology?*
   Most owners I spoke with had at least some experience with tagging animals. The required animal pregnancy scan has also given them some insight into the process of reading a scan. In addition, some of the younger market owners have had experience with computer technology, but only four of the owners I spoke with were under the age of 40 in 2006. Two of those had veterinary medicine degrees, while the others had degrees in animal science. Still, members of the older generation of market owners have almost no computer experience, and those that do have not used that experience to influence the industry as a whole. Most, in 2006, did not even use e-mail, and when discussing the Internet or database technology, one might as well be speaking a foreign language. I should point out that this is not true of every owner. Some auction owners have deliberately sought out the latest technologies, but they are few.

3. *How have traditional beef industry values and current business pressures shaped your opinions about both technology and the NAIS?*
   In general, when asked this question, owners respond that the resistance to technological change comes from their customers more than from

themselves. Most owners reported that their local producers did not want to deal with the NAIS, did not understand it, and were under no immediate pressure to get involved. Furthermore, they said that because there were still so many small producers whose animals make up such a large percentage of the national herd, it would be difficult to get a high degree of participation in any program.

Also, three market owners mentioned that a large portion of small producers are not directly involved in raising animals for large profits, but are interested in the lifestyle that comes with raising cattle. In my experience, many small producers are attracted to raising cattle for the independence it provides and for the outdoor, hands-on work that it involves. It is often a source of extra income and a rural home life that they want. Computers and databases are not a part of that vision even today, for most of these small producers. In addition, the auction owners were quick to point out that their customers are no more literate about technology than themselves. In short, their customers are resistant to change and to the computer technology that might come with change.

4. *Do you feel that proposed new technologies will work within the beef industry?*
   Somewhat surprisingly, more than 75% of market owners answered yes to this question, although they believed that the process would be a lengthy one. Market owners believed that the USDA had done a poor job of defining exactly what those technologies would be and how they would be used. Instead, the USDA, according to market owners, tried to let private industry complete that part of the picture, even though they had not made clear what their own goals were. Market owners also stated unanimously that any technology diffusion would require additional training for their staff. Yet no funds or opportunities for that training had been provided or even discussed. Finally, market owners were clear in their belief that any move to technological means of animal tracking will have to be mandated or demanded by the market before most market owners would be willing to get involved, and even then, they would need much more information about the plan and its regulations.

5. *Does the new business model of the NAIS fit the needs of the beef industry?*
   My sources responded overwhelmingly in the negative to this question. Although they tended to believe that some form of NAIS was inevitable, they also tended to be more than willing to wait for its arrival. Information is a critical commodity in the beef trade. Yet roughly 85% of the market owners I spoke with saw no guarantee in their ability to control their information or to receive data back from higher levels of the supply chain. Packers and feed yards are notorious for their unwillingness to share pricing and purchasing data according. Their reluctance led many small producers to see the NAIS plan as a plan to create one big database that could be used by the "big boys" to manipulate the market to their own advantage.

   Because the USDA was unable to explain to them exactly how data would be controlled and kept private, and because producers and market owners had limited trust in the USDA to begin with, there was little chance that they would be convinced of their data security without specific procedures being laid out and explained to them in detail.

6. *Where have you found most of your information about the NAIS and its accompanying technology?*
Market owners overwhelmingly responded to this question by citing the LMA or another industry alliance, regardless of their background with computer technology or their reported understanding of the NAIS. These alliances are clearly thought of as unions, not just organizations. Market owners believe that the USDA does not have their best interest in mind like the LMA does, although 100% of them report having read at least some of the information distributed by the USDA and other proponents like McDonald's.

Market owners I interviewed told me that information from the USDA was largely unreadable because it is out of touch with their reality. One owner called USDA materials "boring and uninformative." Furthermore, although his customers knew that the USDA materials were available, nobody wanted them because they were not written in a language that they understood.

Producer opinions are a major determinant of market owner opinions in the beef industry. Producers are the market owner's customers. According to market owners, producer opinions are not being formed by the USDA or other NAIS proponents. During my conversations and interviews with market owners, I heard over and over that most cattle producers with no opinion of the NAIS will go to a friend first, a competitor second, and an enemy third before seeking official USDA information on the subject of the NAIS. The result of this avoidance is that most opinions concerning the NAIS are formed "in the coffee shop" as one producer explains his understanding of the program to another. That understanding is almost always based upon an interpretation of the NAIS that has been received from an industry alliance such as the NCBA or a trade journal targeted specifically to the producer's niche within the beef industry.

7. *How would the implementation of the NAIS and the accompanying technology affect your daily business activities and your personal responsibilities?*
The first response to this question was always to cite increased costs and increased demands upon staff and producers. There was genuine resentment among market owners and producers alike toward the USDA for proposing a system that will cost them money while being unwilling to provide funding for training and expenses. Because market owners sell animals on a commission basis, they rely upon their customers' animals to create revenue. And because the producers had little desire to participate in the NAIS, no market owner wanted to be first to adopt the system, thereby increasing costs to his customers and potentially driving them to a competitor.

Second, there was a genuine fear among market owners that any technology that would necessarily accompany NAIS would be cumbersome and difficult for their staff to operate. Every owner I spoke with expressed worry that the NAIS might slow the pace of auctions and drive away buyers as they become tired of spending more hours at the auction and more money on cattle.

8. *Who do you talk to about things like the NAIS, technology, and business concerns?*

According to market owners, there were more meetings on this subject than one might think. But again, the information that most owners trusted came from meetings sponsored by industry alliances and special interest groups. Owners reported that they were much more likely to speak with another market owner or an officer from the LMA or a similar alliance than they are to seek information directly from the USDA or other proponents of the system. It is interesting to note that although market owners are in direct competition with other market owners in many cases, they still prefer their competitors' opinions to more "official" NAIS information sources.

9. *How has the information you have received concerning new technology and the NAIS affected your personal opinion of new technology and the NAIS?*

Not surprisingly, almost all market owners said that their opinion had changed little during the past three years (2003–2006). Most said that their initial interest in NAIS was met with indecision, scarce and confusing information, and poor planning from the USDA. After that, their interest quickly faded. While they had been paying attention since then, owners had not been persuaded by additional information from the USDA or by attempts to explain the good of the NAIS for the beef industry as a whole. If anything, the information flow from industry alliances and other beef industry professionals reduced the willingness of market owners to get involved. Owners typically reported that they trusted the LMA much more than the USDA and that they were unwilling to go against their "union" solely on the basis of USDA propaganda (which they do not trust anyway).

10. *Which information sources were most influential in shaping your current opinion of the NAIS?*

Once again, the clear winners were the industry alliances and peers within the industry. Not one owner told me that his primary source of information concerning the NAIS had been the USDA. Instead, they revealed that information from the USDA strikes them as hard to understand, removed from their personal situation, mired in bureaucratic conventions, and in some cases decidedly against the best interests of owners.

If anything is clear from my conversations with livestock owners, it is that the innovations that would accompany a NAIS, both real and theoretical, are at best misunderstood by many members of the beef community. Rogers [19] describes diffusion in terms of benefits that enhance acceptance of new technologies and ideologies, and possible consequences that limit that acceptance. He also defines innovations in terms of hardware, such as new equipment, practices, and technology, and software, the abstract ideologies whose acceptance requires a deviation from current thought and acceptance of a new paradigm.

1.3.7 Rules from the Road

As shown by my conversations with livestock market owners, the hardware that accompanies the proposed NAIS was still largely a mystery in 2006 even to those who studied NAIS developments. For example, one market owner told me, while I was visiting him in South Dakota, that he had recently purchased a software system that would perform all functions of the proposed NAIS with no additional hardware required. At his invitation, I was able to examine his new software. It was certainly a capable accounting system that could automate certain communication functions, but was in no way capable of creating and tracking the type of data sought by the USDA and the NAIS.

I had many such experiences, but even in the most remote locations I did not find a single market owner who had not at least heard of the NAIS and spent at least some time gathering information about it. Still, two facts were clear. One was that the hardware that would accompany the NAIS was either misunderstood or not understood by most market owners and producers. Market owners believed that the USDA did not understand what will be required in terms of hardware either. Their record of continually changing the plan for a NAIS did nothing to change this assumption.

In effect, then, with no clear description of hardware and its functions to evaluate, market owners and their customers were unable to see the potential advantages of the system.

Even if the requirements had been clearly drawn, they probably would not have understood potential benefits of the system on the basis of the information they received.

The other clear fact was that the top-down communication network that would have maximized the USDA's ability to control the discussion of the NAIS (whereby information would flow from the USDA and other architects of the system to large and small businesses and producers along the beef supply chain) did not exist. Earlier I discussed the notion that most industries have both a formal communication network developed along supply chain lines and also a more informal communication network. In this particular industry, at least on the subject of the NAIS and technology, the informal communication network was the primary communication network. Industry alliances such as the LMA, the NCBA, and R-Calf were much more likely to be primary sources of information than is the USDA. Secondary information sources were most often peers within the industry or others that might be found at local meetings or at the coffee shop. The USDA and other proponents such as McDonald's and other retailers were a distant third in this communication network.

An innovation must have both clear advantages and a means of communication to explain those advantages to would-be adopters in order to gain acceptance. Because the advantages and possible drawbacks of the NAIS were unclear to market owners and their customers, and because communication of its advantages was hampered by unclear and often diverted (through industry alliances) communications, successful diffusion was impossible. The fact that LMA members had the highest change agent status within the livestock market social group, as was routinely reported to me by the owners, did not help the USDA's agenda. As I have shown, the USDA's announcement of the NAIS plan set off a power struggle between industry alliances who sought to control the process of shaping the plan. The ensuing struggle, and the USDA's inability to mediate that struggle, according to market owners, damaged the USDA's efforts to become change agents on the topic of NAIS.

Rogers [19] speaks to this facet of diffusion when he presents *timing* as a crucial element in the diffusion process. Market owners that I spoke with usually told me that the USDA had botched the NAIS rollout from the start. However, the timing of that rollout, combined with poor communications, and failure to recognize the prevalent communication networks within the beef industry were at least as responsible for the ensuing backlash as poor planning, in the eyes of market owners.

Another factor that directly affected the debate over the NAIS is that the plan was not mandated in any way, despite the best efforts of the USDA. The program's voluntary status only added to its continuance as a highly social process. Because market owners turn to peers or friends first for information and advice concerning the NAIS or, for that matter, difficult business decisions in general, communication networks are even more closed and self-reliant than they would be if proponents of the NAIS had more relevant information.

The nature of the beef industry's sociological makeup also contributes to the communication network's influence. Market owners and producers generally fall into the category that Rogers calls "late adopters." People within the beef industry tend to stick to their own kind. They tend to be heavily involved in their work, do not seek influence from external sources, and are highly independent, not overly social, slow to trust and to change, and uninterested in societal fads. This generalization does not apply to all market owners and producers, but few of the owners I spoke with could be described as early adopters based upon Rogers' definition. In fact, most of the people I spoke with that could be described as early adopters were involved either in "higher" levels of the beef supply chain, such as packing or retail distribution, or with a business that produced some sort of technology support for the proposed NAIS.

Market owners and producers also tend to be close-knit, both in their social groups and their communication networks. Their convergence, as described by Rogers and Kincaid [13], happens only within established patterns of communication, between themselves, others like them, and alliances they trust implicitly. They converge only by agreeing to general opinion within their own established communication networks, and they are inherently divergent from external communication networks or even those that are perceived as outsiders. These attributes make them, in Rogers' definition, late-adopters of new technology and ideologies and, in Rogers' and Kincaid's view of communication networks, isolates them within the larger beef supply chain. Therefore, any attempt to persuade them must first conquer the natural defensiveness of the livestock market/producer communication network, and then be extremely persuasive in order to convince a naturally distrustful audience being asked to move out of its comfort zone.

1.3.8   Communication Gaps and Communication Theory

Of course, navigating those communication networks is only part of the equation. As I have said, the content of the message is as important as the process of getting it to recipients. Both must be successful to affect persuasion and technological diffusion. My next course of action was to determine what messages were being given to the market owners and their customers, and to analyze those messages on the basis of my personal communications and observations, communication theory, and linguistic theory. I turn first to my document analysis on the basis of the Elaboration Likelihood Model. I was interested in determining

1. whether an individual message was more likely to be considered using critical thinking skills;
2. the motivation of the reader to consider information presented by the source of the message;
3. the ability of the reader to understand the information being presented by the message;
4. the positive or negative cues most likely to be adopted by the reader in the absence of central route processing;
5. the message's source and bias that may have affected cognition.

The results of my analysis, by document, are recounted below.

1. *Audit, Review, and Compliance (ARC) Branch Policies for USDA Process Verified Program*
   This document was designed to give readers an overview of general policies concerning the USDA PVP and also contains specific procedures for applying for PVP status. The PVP was designed as an accountability tool to allow beef business entities to advertise their adherence to certain quality standards that would ensure product safety. In many ways, the PVP was a marketing tool, designed to recover lost export markets in the wake of the original BSE scare. Proving that certain standards of safety and compliance were being observed would allow, in theory, for companies to market themselves as process veri-fied companies. However, it was the first written product encountered by many American beef industry professionals, which had specific guidelines for quality management. While the PVP is separate from the NAIS, it was a first response to the problems caused by animal disease and inability to trace animals and provide safety assurances within the beef industry. In many ways, it was a "first look" for the livestock market owners and beef producers at the type of bureaucratic requirements that might be presented by a national animal safety system.
      In short, the requirements of the PVP confused market owners and producers. They were not intended to be participants in this program, but because they knew of the impending NAIS initiative, many were interested in the requirements of the PVP. Their fears were quickly confirmed by the most influential sources

within their communication networks. The PVP requirements were designed by the Agricultural Marketing Service (AMS), while the NAIS is directed by the Animal Plant and Health Inspection Service (APHIS). In theory, the two divisions of the USDA have separate missions. This distinction, however, was not clear to many within the market owner/producer communication network.

The communication network members were undoubtedly already biased by the fact that the PVP was a USDA creation. However, many were also intrigued. But the details of the PVP do not lend themselves well to central route processing as defined by ELM. First, the language is intensely bureaucratic. For example, this section gives an overview of the requirements for agricultural product data services:

Data services that do not validate the data being entered into the system are not eligible for approval under the programs.

The ARC branch will allow verification of data services which

1. validate the data entered by the user through on-site evaluations; and

2. validate the data entered by users through data evaluations. Data must be validated to ensure that it is accurate and reliable. Data validation must be addressed within the receiving process under the Programs [36, p. 1].

Using the principles of ELM, it would be almost impossible to conclude that any livestock market owner or producer would be able to use the central processing route to think critically about this passage. The references to data and data services within this section, and throughout the document and its accompanying instructions, only serve to confuse and provide negative cues to the reader. My interactions with market owners and their clients prove that they consider the document vague and ill-defined, requiring the business entity only to provide validation for data. The resulting ambiguity served only to prevent them from understanding, despite their initial motivation, and thus limit central route processing.

2. *ARC 1001 USDA Process Verified Program*
This document was provided by the USDA Audit, Review and Compliance branch as a guideline for applying for PVP status. These are the step-by-step guidelines for applying for the program outlined in the document above. Unfortunately, the specific requirements are often as vague and confusing as the overview. The program requirements cover 19 pages and still vaguely identify specific tasks or procedures to be performed. The document lists these requirements.

The company must

(a) identify the processes needed for the QMS and their application throughout the company;

(b) determine the sequence and interaction of these processes;

(c) determine criteria and methods needed to ensure that both the operation and control of these processes are effective;

(d) ensure the availability of resources and information necessary to support the operation and monitoring of these processes;

(e) monitor, measure, and analyze these processes; and

(f) implement actions necessary to achieve planned results and continual improvement of these processes [36, p. 2].

The individual business was responsible for developing quality standards that will adhere to these standards. Problematically, this approach required those who were already confused about the standards of the PVP to develop their own procedures that would adhere to those standards.

The PVP program requirements continue by demanding the business to produce a quality manual that will meet the following stipulations for a quality management system. Businesses would need

(a) An organizational chart or similar document listing all personnel assigned to managerial positions within the program;

(b) A description of the scope of the QMS, including details of and justification for exclusions;

(c) The specified process verified points;

(d) Documented procedures established for the QMS;

(e) Reference to all forms, tags, and labels used to track or demonstrate product conformance;

(f) A master document list that shows the most current issue of all QMS procedures, forms, tags, and labels used to track or demonstrate conformance;

(g) A description of the interaction between the processes of the QMS; and

(h) All other documentation as required in this Procedure [36, p. 3].

Again, as with the general overview of the PVP, it would be nearly impossible for a livestock market owner to adhere to these guidelines even if they were written in accessible language. They are too vague. It is important to note, once again, that these requirements were not written for livestock markets. They were created for commercial packers and retailers. Yet they were widely distributed among all segments of the beef industry. As with the general overview of the PVP, there is nothing in this document that would allow most readers to use critical thinking skills even if they were motivated and unbiased. Based on my experience with livestock market owners and their customers, bias did exist, and the document's language is too far removed from anything feasible or recognizable to small business owners or cattle producers to counteract that bias through central route processing.

For example, section 1.2.3 in that document addresses document control and reads as follows.

1.2.3 Control of Documents
The company must control all documents required by this Procedure.

(a) A documented procedure must be established to define the controls needed

(b) To control all documents required by this Procedure;

(c) To ensure that changes and the current revision status of documents are identified;

(d) To ensure that relevant versions of applicable documents are available at points of use;

(e) To ensure that documents remain legible and readily identifiable;

(f) To prevent the use of obsolete or unapproved documents; and

(g) To retain all documents for at least 1 year after the year in which the audit was performed [36, p. 3].

While these directives might be usable for someone accustomed to designing audit-based processes on the basis of government directives and the bureaucratic vocabulary, they constitute a foreign language for most members of the livestock industry. The language of the document is far too removed from the daily operation of a livestock market owner or producer to reasonably employ pragmatically based interpretation of the requirements of the PVP for market owners or producers. I intentionally disqualified this document from analysis on the basis of implicature. The document *is* an implicature, having almost no applicable text that can be applied to any daily operations of an average beef industry professional.

3. *The National Animal Identification System (NAIS). Why Animal Identification? Why Now? What First?*

Unlike the previous two documents, this pamphlet was designed specifically for producers and others within the beef industry who might have limited knowledge of the NAIS. It was intended to be persuasive by providing insight into the merits and advantages of the National Animal Identification System. Also, unlike the PVP documents, this pamphlet was published by APHIS, which was directly in charge of the NAIS. Interestingly, the message from the start is that the NAIS is about controlling disease, and the language is much better suited to a livestock industry insider than that used in the PVP documents.

However, on the basis of my interviews and conversations, market owners and producers find more questions unanswered than answered by this document. For example, technology is not mentioned until the final page of the document. When technology is addressed, it is as follows:

> Rather than focus on specific technology, USDA will focus on the design of the identification data system—what information should be collected and when it should be collected and reported. Once the identification system is designed, the market will determine which technologies will be the most appropriate to meet the needs of the system. [30, p. 14]

Although this is interesting and necessary information, it lacks the type of specifics that would allow livestock market owners and others within the beef industry to think critically about the NAIS. This text is readable and understandable because it is not overly reliant on bureaucratic, vague terminology and on technical terminology. However, much like the PVP requirements, it is so devoid of specifics that there is nothing to consider critically. Livestock market owners told me that when they looked at this type of document they were further convinced that the USDA had no plan. Nothing in the document details how supply chain complexities would be handled by the proposed system.

Also, there is no plan outlined to pay for the system or to train personnel to operate whatever technology would be required by NAIS, both of which were clearly central concerns of livestock market owners. When the pamphlet does finally refer to these questions (in the final paragraph) they are addressed as follows:

> Both public and private funding will be required for the NAIS to become fully operational. Database systems must be created and maintained, equipment must be purchased, animals must be identified and tracked, programs must be monitored, and labor is needed for all of these activities. [30, p. 8]

Again, while this information is undoubtedly true, there is nothing to critically evaluate and nothing that would be persuasive based on my interviews with livestock market owners. It could be argued that this document was meant to be merely a brief overview of the program, but most beef professionals saw it as a confirmation of a flawed plan with no direction.

4. *The United States Animal Identification Plan*
This document is a summary of the initial document issued by the NIAA, which was a nonprofit organization that organized a task force of USDA and industry professionals in 2003 to draft a plan for the National Animal Identification System. It is the document that truly started the outcry and debate surrounding the NAIS. The NIAA's inclusion of some groups into drafting the plan at the expense of others created a highly politicized environment surrounding the NAIS. As with the USDA's materials, when viewed through the lens of communication through critical thinking, the document falls short in its ability to engage members of the beef community in persuasive, critical thought.

For example, consider the following passages:

- The infrastructure for individual animal identification will be made available as premises become enrolled to provide for the timely introduction of official ID with the new national numbering system.
- The plan contains no mandatory requirements at this point in its development. Eventually, as the plan is finalized and tested, all livestock and food animals will be able to be tracked through the system.
- While preliminary projections for financial requirements have been made, the plan is still being developed so no specific amounts are yet available [31, p. 7].

The NIAA plan never gets more specific than these general sweeping statements. Like the USDA's brochure, the plan makes grand claims for what will be accomplished by the NAIS without offering any specifics about *how* those goals will be accomplished, what will be required, or who will pay for meeting those requirements. So, while the plan may sound like a good idea in theory, the biased, skeptical readers of the plan that are the producers and livestock market owners were again given nothing concrete to consider and no reason to alter their perception, and were again unable to use the central route processing and critical thinking that would have aided in their persuasion. Even if market owners had been desperate to understand the NAIS, the NIAA's document gave them nothing specific to judge, and no reason to be persuaded that the plan is a necessity.

At the same time, the document makes several promises as to what the plan will do and when its goals will be attained. Among those promises is a call for a national premises registration system to be established in 2004. Also, the plan calls for mandatory participation in the NAIS by 2006. However, over time, both of these target dates and many others came and went. The USDA continued over the next several years to set new deadlines for premises registration and other facets of the program that could not be met. The result was a decrease in the perceived power of the USDA to be sincere about its own policies and to implement those policies. Producers, buyers, and livestock market owners began to question the validity of USDA communications and USDA influence.

5. *Livestock Marketing Association Comments on the US Animal Identification Plan*
   This document is the Livestock Marketing Association's response to the NIAA's report. Despite the fact that the LMA was involved in the drafting of the NIAA's proposal, their objections are evident from the start of their comments. They do not directly attack the NIAA, but immediately turn to commenting on the long histories of livestock markets as focal points for disease control and of livestock market owners in bearing the burdensome costs of those initiatives. The major difference between this document and those produced by the USDA and the NIAA is that the LMA's comments are directly related to the day-to-day operations of the livestock markets themselves. For example, consider the following passage from page 2 of the LMA's response:

> It is important that the cost projections of the ID program as they relate to markets consider much more than just the cost of the "Data Collection Infrastructure" or market readers, if indeed that is what the current United States Animal Identification Plan cost projections cover. For instance, the cost of refitting markets to accommodate the movement of animals through the market and past the readers will likely be much greater than the cost of the readers themselves. Also, we anticipate that the cost of setting up and maintaining a computer infrastructure and hiring technical staff to run and maintain these systems in the markets will be equally enormous. [32, p. 3]

Owners' concerns were based upon financial and training needs, lack of support from and mistrust of the USDA, and lack of specifics from proponents of the NAIS about how the plan would be executed. In the space of one paragraph, the LMA's document does more to address the specific concerns of the livestock market owners than any of the documents from proponents of the system. At the same time, the LMA's response confirms those fears and implicitly invites the membership to oppose the plan by calling the plan's financial cost "enormous."

Furthermore, after examining the lack of funding and underestimated costs projected by the NIAA, the LMA's document goes into an in-depth analysis of the many potential problems they expect livestock markets to encounter on the basis of the vague specifications set forth by the NIAA. The LMA is persuasive in both its ability to speak directly to the concerns of the market owners and its ability to provide a day-to-day operations perspective of the many potential problems with the NIAA plan, largely because the LMA leadership is comprised of livestock market owners. LMA leaders understand the context of other market owners and write directly to their concerns; the commentary is understandable and motivates readers to consider their information carefully, as it relates directly to their daily concerns.

According to the LMA documents, the NIAA plan claims to offer the ability to track livestock. Yet, as the LMA points out, there are innumerable variables and animal comingling variations that must be addressed before any such plan could even begin to become reality. The USDA describes no plan for those eventualities in its documents. USDA plans have no procedures for moving animals through tagging at livestock markets, what information would be required at the markets, or who would oversee the proper coordination and compliance of those activities.

6. *National Cattlemen's Beef Association NAIS Industry Proposal White Paper*
This white paper was the response of the National Cattlemen's Beef Association to the NIAA report. Like the LMA's response, the NCBA's paper immediately focuses upon the financial concerns of its membership, saying that the USDA has yet to receive the necessary level of funding to make the NIAS fully implemented, and thus, like the LMA's document, speaks directly to one of the primary concerns of its membership.

However, unlike the LMA's document, the NCBA goes on to provide a rationale for a completely private system, also by speaking directly to constituent concerns over data privacy. Because many producers are greatly concerned over the federal government or large businesses gaining access to their personal data, the NCBA's proposal calls for data service providers and data trustees, both of which would be private industry sources, to control the data and to provide services only in accordance with mandates approved by the NCBA's constituents and the technology firms serving as contractors to the privatized national system [33].

Interestingly, the NCBA does not discount the need for NAIS, but the document is persuasive to its membership because of its ability to speak directly to member concerns over price fixing by large companies, unfair competition, and the USDA's inability to describe how confidential data will be protected. Potential bias can have a strong influence upon persuasion. The NCBA proposed a privately held plan that would be more in tune with the desires of its membership.

Yet it is highly persuasive in its promises to privatize the system, to remove "Big Brother" from the equation, and to protect system data from outsiders. The NCBA's promises have an undercurrent that says, "We know you don't understand, but we do"—more so than the LMA's direct interpretation of the effects of any NAIS. Still, the promises provide comfort to the NCBA audience by removing the responsibility of compliance from the average producer, which correlates with one of the postulates of ELM. By removing personal responsibility from the audience, the NCBA document, in many ways, encourages members to use the peripheral ELM route when considering NAIS and to defer to the NCBA "union" on the matter.

7. *R-Calf USA 2006 Position Paper: National Animal Identification System*
R-Calf's response to the proposals that came from the NIAA's meetings is straightforward. Unlike the LMA and the NCBA, the concerns of the R-Calf organization, which consists mostly of cow-calf producers, center upon whether a need for a national animal identification system truly exists. R-Calf initially proposed country-of-origin labeling, which would have labeled each package of meat according to its country of origin. In its view, the need for a national animal identification system has never been truly demonstrated. It believed that labeling beef products by country of origin would give consumers the information needed to strictly regulate quality within the US beef supply chain.

Yet, like the documents from the NCBA and the LMA, R-Calf's documents consistently speak to its members' concerns, and like the LMA document, to the specific concerns of their members that are related to daily business operations. R-Calf's position is that any system must remain in public, rather than private hands, because of a private system's potential for abuse by large business interests.

This position is consistent with R-Calf's commitment to the "small operator" [34] and traditional values such as maintaining producer freedom and a level playing field with large operators. For example, R-Calf believes that any national system must include traditional methods of animal identification such as branding. Traditional identification methods would be difficult to incorporate into the digital system proposed by the USDA and other alliances. However, like the responses from the LMA and the NCBA, R-Calf's commitment to branding, public funding, and data protection is consistent with the values of its members, as I found in my discussion with many northern state (where R-Calf is most prevalent) producers and livestock market owners.

Once again, on the basis of those values, R-Calf's positions can be much more easily analyzed by its members than those of the USDA because of the language used, the paper's ability to link the issues to producers' daily concerns, and its ability to speak to the fears concerning the NAIS that hamper the USDA's efforts among producers.

Also, like the NCBA and the LMA, this paper was produced by an organization with a singular mission to safeguard the livelihood of its members. The members of R-Calf would be more motivated to read this paper than one from the USDA on the basis of this fact alone. But because the content of that paper is consistent with the daily operations of its membership, as opposed to the USDA's publications, the reader's motivation and ability to process the information are greatly enhanced. Hence, the readership's understanding of the NAIS is much more likely to be drawn from the R-Calf documents than from documents produced by the USDA.

The message of R-Calf to its members is more prohibitive than inclusive. Unlike the messages of the LMA and the NCBA to their memberships, R-Calf's document seems only to indicate its willingness to fight against a mandated system and to guard small producers' data against an incursion by the federal government. R-Calf was direct in its questioning of the NAIS and has generally refrained from giving its membership directives. Still, as with the other industry alliance documents, R-Calf is much closer to the concerns of the average reader than anything produced by the USDA.

1.3.9   Textual Analysis with Implicature and Pragmatics

As you might expect from the results shown by analysis with communication theory, the results of my investigation show large differences in the numbers of implicatures and in the extent and depth of pragmatic consideration given to readers in the definition of context and background knowledge. While none of the texts overtly describe the type of background knowledge required by the reader, it is easy to see that the texts are written for very different audiences in terms of their background, and the implicatures required of readers show a clear pattern.

In general, the USDA and NIAA documents show a consistent disregard for both reader vocabulary and reader background knowledge. Although there are a handful of beef supply chain professionals that would understand the USDA's documents, they are far from the norm. Most members of the supply chain expressed to me in conversations and interviews that the type of terminology found in the USDA documents is largely foreign to them and to their customers. Furthermore, when implicature is required by the documents from the USDA and the NIAA, it is consistently reliant upon ideologies far removed from the daily activities of any market owner or producer. Finally, and this is perhaps the most pronounced difference, the texts from the LMA, NCBA, and R-Calf only introduce one concept at a time and tend to base their analysis of any given concept from the USDA and NIAA documents upon "real-world" concepts that can be pragmatically applied by most of their members. The difference in the number of counted implicatures required of readers is largely a result of this. Total implicature counts for the seven documents analyzed and the terms that required implicatures are documented in Tables 1.1 and 1.2.

The implicature count results per document are somewhat misleading because of their differences in referring expressions. As previously noted, each of the documents seemed to have a different persuasive purpose. It is tempting, looking at the implicature

TABLE 1.1. Implicatures required per analyzed text are shown.

Text	Implicatures Required
Audit, Review, and Compliance (ARC) Branch Policies for USDA Process Verified Program	14
ARC 1001 USDA Process Verified Program	Disqualified because of overwhelming implicature
The National Animal Identification System (NAIS). Why Animal Identification? Why Now? What First?	10
The United States Animal Identification Plan General Overview	8
Livestock Marketing Association Comments on the US Animal Identification Plan	4
National Cattlemen's Beef Association NAIS Industry Proposal White Paper	25
R-Calf USA 2006 Position Paper: National Animal Identification System	0

count results, to classify the NCBA's document as the most poorly written. But, as I noted earlier, the NCBA document seemed from the start to be aimed at removing the NCBA constituents from the process of implementing the NAIS.

Overall, per page, the USDA and NIAA documents exhibit the most implicature requirements of the reader, especially given the complete lack of pragmatic consideration given to each. The documents prepared by the LMA, NCBA, and R-Calf are constructed for specific purposes with specific audience concerns in mind. They also show a direct correlation to the daily activities (with the exception of most of the NCBA document) to members' daily activities. They are documents constructed for a purpose and addressed to US beef professionals. This is more than can be said for the USDA and NIAA documents, which are far too vague and far too removed both linguistically and practically to be either communicative or persuasive for most of the beef industry audience.

1.4 Lessons of Beef and Bandwidth

Considering my results, there was clearly misunderstanding and disagreement between the proponents of NAIS and its technology and the livestock market owners and producers who would need to adapt to the NAIS and its technology. I began to notice the gap between the two factions almost as soon as I started talking with cattle producers and livestock market owners. Even the coffee shop talk that was going on in livestock markets made it obvious that there was some level of disagreement and misunderstanding. Of course, that did not explain what was causing the rift. I believe that the misunderstandings can be traced to poor communication strategies employed by NAIS proponents, coupled with a lack of systematic planning and pragmatic consideration.

TABLE 1.2. These terms required implicatures of readers.

Text	Implicatures
Audit, Review, and Compliance (ARC) *Branch Policies for USDA Process* *Verified Program*	Verification Data services Validate Data Program Data System Non-validated data On-site evaluation Data evaluation Data validation USDA Process Verified Program Audit Principles of auditing ISO 19011:2002 Independent and systematic Evidence-based approach
ARC 1001 USDA Process Verified *Program*	Disqualified on basis of near complete implicature
The National Animal Identification *System (NAIS). Why Animal* *Identification? Why Now? What First?*	Identification data system National Data Standards System architecture Premises and animal identification State or tribal animal health authority Animal identification number distributor Database Zoonotic disease outbreak Identification system Design of the identification system
The United States Animal Identification *Plan*	Radio frequency technology Individual animal identification ISO 11784 ISO 11785 Code structure Technical concepts Code of federal regulations Official identification devices
Livestock Marketing Association *Comments on the US Animal* *Identification Plan*	Canadian ID experience Non-producer participant Market participant RFID tags warrant Data file transmission

TABLE 1.2. (*Continued*)

Text	Implicatures
National Cattlemen's Beef Association	NAIS architecture
NAIS Industry Proposal White Paper	Centralized database
	Service provider
	Data Trustee
	Private database
	Information management company
	Animal health authority
	Robust technical solution
	NAIS network
	NAIS database
	Data collection
	ISO
	HDX
	FDXB
	Mobile applications
	Ramping available data
	Interfacing
	PDA
	Import identification numbers
	Hosting facility
	Redundant hardware
	Disaster recovery
	Off-site storage
	Middleware data-scrub software
	Merge and search application software
R-Calf USA 2006 Position Paper:	None
National Animal Identification System	

1.4.1 No Pardon for Jargon

Written documents highlight differences in the views espoused by the USDA and other proponents of NAIS compared to the views of NAIS opponents. In general, the materials from the USDA were much more technical, had a higher incidence of implicature, paid less attention to background knowledge and current industry trends, were not targeted to key stakeholders, and largely ignored communication networks. USDA documents were also less likely to target themes that promoted central route processing and critical thought. In short, the USDA's persuasive strategies and methods of technical communication were ineffective because of their lack of audience perspective, their avoidance of livestock market owner and producer business concerns, and their inability to use communication networks to successfully diffuse NAIS technology. In hindsight, I believe that the USDA would have had a hard time selling NAIS to the beef community even if it had produced solid, pragmatically considered materials, because the USDA alienated primary beef industry communication network change agents, such as the LMA and

NCBA, from the start. But, misunderstandings only served to compound that error and to prevent any recovery.

It seemed clear from my initial conversations and observations with market owners, producers, buyers, and others that there are a wide variety of skill sets and expectations within the beef industry concerning NAIS and its technology. To change technology resistance in this case, it was imperative to promote that technology in language that made sense to the stakeholders involved.

1.4.2   Alice Is Not in Wonderland

The office managers, who are, as a group, the most technologically advanced employees at livestock markets, seemed willing to entertain the idea of new technology and new ideologies, but they were so unfamiliar with the types of advanced technology required of them by NAIS that they would still require training, as would their staff members. Also, they were, in general, unwilling to take on the role of tech-support person in addition to their already busy schedules. The market owners knew less than the office managers about computer technology, and in general did not want to know more. Of course, the office managers were not targeted for persuasion or education on the facts of NAIS like the owners were. So, in effect, the most technologically knowledgeable personnel in the livestock market industry were ignored by NAIS proponents. It is true, of course, that the level of technological sophistication among office managers was not generally sufficient to maintain NAIS systems at livestock markets on a national level, but they would probably be a more fertile starting ground for exploring technological possibilities than the market owners.

Also, there is a prevalent sense of disdain for technology among many producers. Younger producers are much more accepting, but there are fewer of them. Older producers are much more rigid in their ways. As a group, they do not trust technology. Many of them are of an older generation, which means that they did not grow up with computers or software. In fact, they have a proud sense of self-reliance that leads them to frown upon new ideologies and methodologies and to adhere to the old ways of doing things. Producers are individuals who are accustomed to working with their hands and are not comfortable with processes that cannot be seen or understood. Their distrust of the federal government is strong, and that distrust is particularly true of the USDA.

Poor communication from NAIS proponents did nothing to alleviate their concerns. If anything, those communications caused more confusion. That confusion was, in many ways, the beginning of resistance to the NAIS among livestock market owners and their customers. When confronted with a controversial innovation, they were forced to consider the innovation and to attempt to understand the benefits and drawbacks of the NAIS. Yet they were unable to gather the information they needed to make an informed, critical decision.

Most market owners indicated that they had, at least at one time, been intrigued by the idea of NAIS. However, over time, most of the owners soured on the idea, and many quit paying much attention to the latest developments. In addition, most owners that I spoke with profess no love for technology in general. It is not that they do not respect technology's potential, but they see technology in terms of time and priorities.

Most owners put in long work days, and learning new computer skills was definitely at the bottom of their priority list. They saw the NAIS and new technology as a ticket to more work and less profit. Communications about the NAIS in its infancy did nothing to alleviate these central concerns.

1.4.3   The Telephone Game Still Happens

Information is shaped as it travels. Communication networks had a huge impact on early acceptance/rejection of the NAIS. Any form of the NAIS that would be able to acceptably diffuse throughout the beef industry will need the backing of the industry alliances and the willingness of the communication network as a whole to pass along accurate and understandable information. The alliances, in conjunction with influential individuals throughout the industry, are the true conduits of information and bridge the distance between government and corporate entities and the thousands of small business owners spread around the country.

Unlike a manufacturing business, for example, members of the beef industry are widely dispersed. There are cattle producers and livestock markets all over the country and they reside in predominantly rural areas. The distances between members of the communication network directly affect personal communication and even further empower the effects of coffee shop diplomacy and establish communication networks. There are parts of North and South Dakota where a person can drive on a state highway for half an hour without seeing another car. Communications in these rural areas take place between trusted friends at regular meeting places. Communications are personal and influential. This is not true for the entire industry, but Web-based communication is certainly not the dominating force that it is in many industries. So, even though the USDA's website contained a plethora of information about the NAIS, many members of the beef industry never saw that information. Instead, as I have shown, they relied on interpretations from others within their communication network. Thus, market owners and producers, in particular, were left to form their own opinions on the basis of the information that they had.

The market owners, for their part, believed that the number of small producers is decreasing continuously anyway, as fewer young men and women want to get involved in raising livestock because of shrinking profit margins and urbanization. The NAIS, in their opinions, would only have further consolidated production in the hands of "Big Business." Over time, many continue to believe that consolidation will take place, resulting in fewer producers, more large production ranches, and increased opportunities for a program like NAIS to become feasible. But, they do not want this outcome, and despite the efforts of the USDA and companies like McDonald's, most market owners and producers were not persuaded of the need for the NAIS.

Many producers and market owners, who are of the pre-computer age, candidly told me that they hoped to retire before the industry took on new digital ways of doing business and transferring information. They saw it as dehumanizing to their business and as an affront to their traditions that hold a handshake as a firm commitment, regardless of contracts or database figures.

1.4.4   It All Comes Down to Doin' Business

There were also practical matters that were not fully thought through by NAIS proponents. At certain times of the year, a weekly auction can last 12 hours or more (sometimes more than one day). A small delay in processing each animal sold at auctions
could greatly increase sale time, which *nobody* wants. To be fair, the USDA promoted
premises identification (and still does), which would allow animals to be largely identified before arriving at auctions. This could eliminate problems caused onsite at livestock
markets. Also, many livestock markets allow animals to be dropped off the day before a
sale, which would also cut down on sale delays. However, animals are often paired with
other animals before, during, or after a sale, which would make tracking more difficult.

In fact, one recent study found that of two million animals studied in Colorado,
almost 90% had comingled with animals identified as "sick" by the research team at
some level [37]. Cattle tend to move often once they begin to be sold, which leads to
high levels of interactions with other animals. The same study found that all 19,000
identified premises in the study had had some level of contact with the infected animal.
Does this mean that all 19,000 premises would be infected by a communicable disease?
No. But the ability to trace an animal with computer hardware and software does not
necessarily guarantee containment of a communicable infection, which was another sore
point among producers and market owners.

Still, it is not hard to see that the ability to trace animals would be of value in terms
of tracking disease. As I have said, several countries such as the United Kingdom and
Australia are already operating successful tracking programs. But those programs are
mandatory in most cases and had significant government financing in their infancy. They
also enjoy a level of support from members of their beef industries that has yet to be
seen in this country.

The same technology that would drive the NAIS could also be used, at least in
theory, to relay information from packers to producers and feed lots that could be very
valuable and could potentially lead to better animals and better products for consumers.
But that relies on the assumption that members of the supply chain would be willing
to share information and that the technology itself could be managed by those who
would benefit.

Technology would be a pivotal factor in successfully selling animals and tracking
data within any form of NAIS plan. Therefore, any NAIS plan and the technology that
would accompany it would need to be completely understood by members of the supply
chain (especially among members such as producers and livestock market owners) and
would need to operate both efficiently and quickly. Because the USDA, in the opinion
of market owners and producers alike, failed to explain how that would happen, they
were distrustful and resistant to the ideology and technology of the NAIS.

The advantages of new technologies and ideologies cannot be confused with the
social processes that lead to their acceptance. Even when an idea has merit, as we
have seen, communication networks, background knowledge, experience, and social
dynamics all directly affect the success of technology diffusion. That is exactly what
has happened with the NAIS. The USDA relied upon the technological efficiency and
safety concerns to sell its ideology, while ignoring the societal values and technological

abilities of their audience. They also failed to communicate with members of the beef supply chain in a manner that could have overridden their concerns with clear, relevant information about their plans.

Within the highly competitive business environment of livestock market owners and their customers, this type of ambiguity is interpreted as a lack of planning and is distrusted. Financing is a central concern of the market owners and their customers. The statements made by USDA documents and communications did nothing to dissuade their fear of being required to pay for the program themselves and for training their employees. Therefore, rather than engaging them in critical thought, these partially considered communications lead to an emotional response based upon long-standing distrust of the USDA and fears of overwhelming changes to business models, technology, and costs.

Many members of the beef community that I interviewed never got past the initial price tag of the NAIS. However, some did, and were genuinely interested in understanding the technology that would support such a bold initiative. On that account, my interviews and conversations with livestock market owners and other beef industry professionals revealed several consistent themes:

- First, livestock market owners and their customers are clearly more influenced by industry alliances than by the USDA.
- Second, although many made an effort to read and comprehend the plans laid out by the USDA, the industry as a whole found those materials to be detached, confusing, and incomplete.
- Third, the portions of those materials that were understood by market owners, their customers, and industry alliance officers were not persuasive.
- Last, that personal contacts and informal communication networks are still vital to communication within the beef industry.

As I have shown, one of the USDA's first actions, although unintentional, was to create a virtual "turf war" among the predominant industry alliances that removed the USDA from the role of change agent and cast it in the role of "Big Brother." Materials distributed by proponents of the NAIS were system-based rather than user-centered. Those materials did not create a common background between writer and reader, which was confusing and disheartening for members of the beef industry.

Thus, fearing for their own livelihoods, and convinced that the USDA's "mystery plan" would either enslave them or bankrupt them (or both), the individual alliances, which are, as I have also shown, the overseers of beef industry communication networks, began a fierce anti-NAIS campaign designed to wrest control of the NAIS from the USDA and protect their own interests. Had they access to a fully realized plan from the start, I do not believe that their reaction would have been so antagonistic.

Furthermore, the USDA responded to the initial industry backlash against the NAIS by promoting a string of "deadlines" for the adoption of NAIS, all of which were later postponed. Over time, these unrealized deadlines were treated much the same as any empty promise—they were ignored.

Even if the USDA had moved past pricing and politics before the debate over the NAIS began, it still would have needed better communication strategies to sell the idea. As my results show, the USDA's written communications suffered dramatically, as did its informal communications, as a result of its divergence from communication maxims that could have helped channel key communications through social networks and better defined written communications for intended audiences. Grice's model shows that we expect an appropriate amount of information from communication and that we expect it to be relevant. Neither of those conditions was met by early communications about the NAIS.

1.4.5 What We Have Here Is a Failure to Communicate

Motivation was not originally a problem for NAIS customers. Market owners saw themselves as responsible for performing specific tasks related to NAIS and for adhering to NAIS policy. They were, therefore, sufficiently motivated to form correct opinions and behaviors with respect to the NAIS. We can also assume that they were not initially distracted, because all information about NAIS was initially coming from the USDA. For the same reason, there was no initial overabundance of information about NAIS. The overwhelming number of messages and message sources concerning NAIS presented themselves only after the industry had determined that the USDA plan was incoherent and industry alliances began to vie for control of the process. Afterward, old resentments and distrusts began to make NAIS diffusion both social and political.

The ELM states that new information encountered by message recipients with an existing bias toward the source of the message or the subject will likely bolster the existing attitudes of the message recipient. However, as ELM also points out, the personal relevance of a message to the recipient can overcome this initial bias, leading the recipient to attempt to form objective opinions in spite of original bias. The NAIS, as a primary business concern for livestock market owners and producers, provided plenty of motivation for them to understand its principles. For that reason, we cannot conclude that bias was an initial impediment to the central processing route for them, even though bias did exist. We can also conclude that there were no initial distractive communications that would have prevented them from understanding NAIS. And finally, we can conclude that the livestock market owners and producers, as nonoriginators of the plan, were shielded from personal responsibility for its success.

Thus, having eliminated distraction, motivation, bias, relevance, and personal responsibility as factors that would prohibit communication, we must place the blame on the communications themselves. The only other factor that blocks critical thinking, persuasion, and information processing is a message recipient's inability to understand the content of the message. We have seen that most market owners were at one time motivated to understand NAIS. The only missing link was their ability to understand the information presented. My results show that the information presented to market owners by the USDA was not understood because that information was not designed for the beef audience, and when it was understood, it was so pragmatically removed from the daily operations of beef industry professionals that the information did not speak to their concerns. That translates into *no plan*, and no concern for those impacted by the plan. Hence, emotional responses and communication network sources (such as the industry alliances) became dominant, and the principle determinants of technology diffusion success became external interpretations of NAIS policy from sources claiming to represent the message recipient's best interests.

Yet, there were also problems with USDA communications at a textual level.

> Even if the documents produced by the USDA and other NAIS proponents had been on topic and pragmatically centered, they were often so poorly written that they could not have been understood or persuasive.

Pragmatically, the documents seem to be created for no one in particular and by no one in particular, and they contain very few specifics about anything in the context of daily beef operations.

Written communication relies upon creating successful implicatures through phrasing and terminology. We expect enough relevant information, based on shared knowledge, to communicate without explanation of every facet of a discussion. The consistent need of the USDA and NIAA's texts for extensive implicature seriously hampers the ability of the text to ease beef industry readers into NAIS understanding. By consistently referring to concepts unavailable to the reader, they create confusing passages without explanation. As soon as the reader is making progress, the reader finds another term that requires implicature based upon unavailable references.

In contrast, the documents from the industry alliances (with the exception of the NCBA) and newspaper and magazine articles (which tend to oppose the NAIS) typically reference familiar situations and terminology. The bureaucratic terminology employed by the USDA to streamline discussions is a victim of its own practices. The very terminology employed in its writing style is misunderstood by its customers. It requires implicatures through references, obscure terms, and concepts that do not lend themselves well to everyday beef industry practices.

Implicature is essential to both the reader and the writer, because it decreases our need to explain every little thing. But, it is also essential that implicatures be built upon common background knowledge shared by both the reader and the writer. In the case of NAIS proponents and the beef industry, this shared knowledge was not properly incorporated into communications (especially documents). When we speak of specialized concepts such as the NAIS and technology, implicature becomes more of a burden than it does in regular text. Documents promoting the NAIS clearly fall into this category. Because those documents frequently refer to specialized concepts, as opposed to everyday concepts, implicatures are more frequently required and are much more difficult for the reader to create. In that situation, the burden falls upon the writer to use text to create a situational model that encompasses a delineated starting point. In this case it did not.

In short, it is a long way from a desk at the USDA's Washington, DC, offices to a sale barn in Texas. If the writer does not share background knowledge with the reader, then the writer and reader cannot share pragmatic meaning of language. The situational context of USDA directives, written from a system-centered point of view by USDA employees, is too far removed from the context of anything happening at that Texas livestock market to be integrated into everyday business.

If we return to my initial examples of implicature concerning the couple going out to eat, we can see that they are built on everyday concepts such as money, restaurants, and the link between the two. This correlation does not appear within NAIS discussions from the USDA, where a reference to something like a *database* may have no inherent connotations whatsoever for the average reader, unless, of course, we know the context from which the reference came, such as the setting of the couple being stranded in a raft, or in the case of the NAIS, the context of the livestock market. Again, market owners and producers were not given that context.

The USDA, and others trying to explain the merits and everyday workings of the NAIS to market owners and producers, would have been better served by creating hypothetical situations, such as that of a market owner who wanted to implement NAIS and has concerns. The most common scenarios could have been identified and used in a way that would have provided at least some type of contextual backdrop and less confusion.

Pragmatic consideration should also be given to similar efforts in the future. How can a producer be expected to understand the complexities of a national system unless those complexities are first explained in terms of a small ranch? Regardless of the specific communication, the need to include implicature, pragmatics, communication theory, and core industry values into similar programs is real. They are also fundamental to any industry that seeks to implement radically new and different technology, and the beef industry is certainly not the only industry that will be asked to adopt radically new technologies in coming years.

The NAIS plan sought to expand its prerogative to almost every agricultural industry, and similar technologies are poised to radically alter manufacturing, shipping, and other agricultural industries, just to name a few. Those industries will face communication challenges similar to those that the beef industry has experienced.

Unfortunately, current technologies seem to change before our understanding of them is complete. Technology will continue to change, computer technology will continue to affect new industries, and there will be a need for extensive guidance and education concerning that technology.

The overseers of the beef industry communication networks, such as the LMA and NCBA, have considerable political clout. They are seen as "grass roots" organizations in their home states, and politicians are reluctant to offend them, especially in highly agricultural states. In fact, most politicians I spoke with were interested in promoting animal identification only as far as it would be supported by their constituents; and even when support is high in a certain area, supporting the initiative with state or federal dollars seems to strike them as a risky proposition.

1.4.6 Culture Is King

Several themes were apparent during my investigation into the beef industry and seem likely to resurface during similar struggles in the future. The first of these was that beef industry professionals see what is going on in the world through trade journals and magazines, from reports from the industry alliances, and even through reports from the USDA. They understand that global marketing is important for their future and that other countries are making concerted efforts to promote their beef as safe and traceable. They also understand that the beef market may quickly become reliant upon traceability. But, they were unwilling to commit their current financial standing to a program that was been poorly communicated, is ill-defined, and had no political backing.

Market owners and producers are not stupid. Cattle owners know cattle. Many of them are very successful financially. Their financial standings and lifestyles make them independent in a way that most of us cannot understand in today's world. But the convergence of cattle and computer technology is just starting, and will be difficult. If we took computer programmers and placed them in the back of a cattle auction, asking them only to move cattle from the pen to the sale arena, there would be pandemonium!

The NAIS, as a plan, existed only as a construct to be argued for or against on the basis of limited information. As a group, the beef community *is* willing to use new technology, although they insist that it be understandable, beneficial, and profitable. In an industry like the beef industry it stands to reason that digitizing some or all of their processes could save both time and labor. This concept is something that all participants seem to support. To them, the only reason to use technology is if it makes things easier and more profitable. Confusion among any technologically inexperienced group of users is met with immediate frustration and must be countered by effective explanation and pragmatically centered design.

Finally, the skill set likely to be encountered within this industry and others like it that are new to computer technology is prone to be widely varied. While not common anymore, it is still possible to operate a livestock auction without a computer at all. As we have seen, there are some who have no intention of having anything to do with hardware or software or anything else computer-related. In turn, there are those who we would not expect to be savvy, who are. This must lead us to consider the needs of the beef industry user from an all-encompassing perspective, while realizing that each individual user expects a certain amount of tailoring to his needs.

One thing that all participants seem to realize is that the industry is changing. Whether they want to be involved or not, they all seem to realize that eventually *someone* will *have* to be involved.

Whatever the future holds, it is my hope that the findings of this investigation will prove to be beneficial in the pursuit of a better technological future. For that matter, I hope that the model employed here will be beneficial to other industries as well. I believe that the combined analysis of technology diffusion, communication networks, theories of communication, and principles of written communication can be adapted to other industries that are moving into more technologically driven modes of operation.

Technology is advancing at such a rate that we cannot hope to keep up, save for our own limited areas of expertise. That means that entire industries will need communicative strategies for explaining computerized technology to new users. That situation is in no way limited to the beef industry.

1.4.7 The Situation Now

As of 2013, the USDA announced its final rules for animal identification [38] within the United States. The rules that will be implemented are a far cry from those envisioned in 2004 and even as late as 2009. Most of the original plans proposed by the USDA have been abandoned. For example, identification of any kind applies only to animals that move across state lines. That identification can be any one of several low-tech documents including veterinary inspection, owner-shipper statements, or brand certificates. In addition, animals going to slaughter can be identified with only a back tag, and brands, tattoos, and breed registry certificates are now considered official forms of identification. Finally, only animals over 18 months of age are subject to any identification requirements, and electronic identification, centralized databases, and movement reports have been eliminated completely. In short, very little has changed.

References

1. S. Stecklow, "U.S. falls behind in tracking cattle to control disease," *Wall Street Journal*, Jun 21, 2006 [Online]. Available: http://online.wsj.com/news/articles/SB115085603103785984

2. I. Convery et al., "Death in the wrong place? Emotional geographies of the UK 2001 foot and mouth disease epidemic," *Journal of Rural Studies*, vol. 21, no. 1, pp. 99–109, 2005.

3. A. El. Amin. "BSE in Canadian cow sets back efforts to regain lost markets," *Food USA*, Jan. 24, 2006 [Online]. Available: http://www.foodproductiondaily.com/Safety-Regulation/BSE-in-Canadian-cow-sets-back-effort-to-regain-lost-markets.

4. J. Andrews, "Imports and Exports: The global beef trade," *Food Safety News*, Nov 18, 2013 [Online]. Available: http://www.foodsafetynews.com/2013/11/imports-and-exports-the-beef-trade/.

5. J. Lawrence and D. Otto. "Importance of the United States cattle industry," *National Cattleman's Beef Association*, Mar. 12, 2005. Available: http://www.beef.org/NEWSEconomicImportanceOfTheUnitedStatesCattleIndustry2704.aspx.

6. *Beef Industry Statistics*, Beef USA. Available: http://www.beefusa.org/beefindustrystatistics.aspx.

7. *Directory of Service Providers*, Drovers Cattle Network. Available: http://www.cattlenetwork.com

8. B. Scheiler. "Tech firms stampede to livestock," *Wall Street Journal*, p. B.3C, August 24, 2005.

9. D. McClelland. "McDonald's exec: BSE testing adequate, ID needed," *Farm Week*, Jun 10, 2005 [Online]. Available: http://traceability.blogspot.com/2005/06/mcdonalds-exec-bse-testing-adequate-id.html.

10. S. Clapp. "USDA releases guidance on animal ID devices," *Food Chemical News*, March 13, 2006 [Online]. Available: http://www.informa.com/.

11. S. Clapp, "USDA considers new animal ID approach," *Food Chemical News*, vol. 47, no. 49, p. 9, 2006.

12. J. Gilmore, "Need-to-Know info on animal ID," *The New American*, May 15, 2006 [Online]. Available: http://www.highbeam.com/doc/1G1-146271605.html

13. E.M. Rogers and D.L. Kincaid, *Communication Networks: Toward a new paradigm for research*. New York, NY, USA: Free Press, 1981.

14. R. Paul, "Stop the national animal identification system," *Lew Rockwell Online*, May 29, 2006 Available: http://www.lewrockwell.com/2006/05/ron-paul/stop-the-nais/

15. M. Zanoni, "The national identification system: A new threat to rural freedom?" *Country-side and Small Stock Journal*, Jan/Feb, 2006. Available: http://www.countrysidemag.com/90-1/mary_zanoni/. Accessed on January 20, 2008.

16. "Senator Talent introduces measure to prohibit mandatory animal identification," *Project Vote Smart*, Sept 7, 2006. Available: http://votesmart.org/public-statement/207026/#.U4OGGihurFA.

17. J. Roybal, "Our crystal ball," *Beef*, Oct 1, 2006 [Online]. Available: http://beefmagazine.com/mag/beef_crystal_ball.

18. P. Shinn, "Congress slashes animal ID funding," *Brownfield Network*, Dec 24, 2007. Available: http://old.brownfieldagnews.com/gestalt/go.cfm?objectid=0DC66BA6-CD91-5F27-28E035095FB1762C.

19. E.M. Rogers, *Diffusion of Innovations*. New York, NY, USA: Free Press, 1995.

20. D.M. Fetterman, *Ethnography: Step by Step*. Thousand Oaks, CA, USA: Sage, 1998.

21. J. Van Maanen, *Tales of the Field: On Writing Ethnography*. Chicago, IL, USA: University of Chicago Press, 1988.

22. R.E. Petty and J.T. Cacciopo, *Communication and Persuasion: Central and Peripheral Routes to Attitude Change*. New York, NY, USA: Springer-Verlag, 1986.

23. P. Grice, "Logic and conversation," in *Syntax and Semantics*, Vol. 3: (Speech Acts). New York, NY, USA: Academic Press, 1975, pp. 41–58.

24. D. Blakemore, *Understanding Utterances: An Introduction to Pragmatics*. Oxford, U.K.: Blackwell Publishers, 1986.

25. D. Schiffrin, *Approaches to Discourse: Language as Social Interaction*. New York, NY, USA: Wiley-Blackwell, 1994.

26. R.A. Zwaan and M. Singer, "Text comprehension," in *Handbook of Discourse Processes*. Mahwah, NJ, USA: Lawrence Earlbaum Associates, Inc., 2003, pp. 129–145.

27. J. Van Maanen, "An end to innocence: The ethnography of ethnography," in *Representation in Ethnography*. Thousand Oaks, CA, USA: Sage, 1995, pp. 1–35.

28. United States Department of Agriculture, "Audit, Review, and Compliance (ARC) Branch Policies for USDA Process Verified Program and USDA Quality System Assessment Program," 2006. Available: http://www.ams.usda.gov/AMSv1.0/getfile?dDocName=STELDEV3103483. Accessed on June 26, 2015.

29. United States Department of Agriculture, *USDA Process Verified Program*, 2006. Available: http://www.ams.usda.gov/AMSv1.0/getfile?dDocName=STELDEV3103489. Accessed on June 26, 2015.

30. USDA Animal and Plant Health Inspection Service, *The National Animal Identification System (NAIS): Why Animal Identification? Why Now? What First?*, October, 2004. Available: http://www.aamp.com/documents/APHIS-AnimalIDBrochure_000.pdf

31. National Institute for Animal Agriculture. *The United States Animal Identification Plan*, January 15, 2004. Available: http://www.animalagriculture.org/Information/Hot%20Topics/Animal%20ID/USAIP%20Handout.pdf

32. Livestock Marketing Association, *Comments on the US Animal Identification Plan*, January 26, 2004. Available: www.lmaweb.com.

33. National Cattleman's Beef Association. *National Animal Identification System (NAIS): Industry Proposal White Paper for Consideration*, December 24, 2005. Available: https://admin.beef.org/udocs/NationalAnimalIdentificationSystemFeb2005.pdf

34. R-Calf USA, *R-Calf USA 2006 Position Paper: United States Animal Identification System*, January 24, 2006. Available: http://r-calfusa.com/Animal%20ID/06PositionPaperAnimalID.pdf

35. R.W. Gibbs, "Nonliteral speech acts in text and discourse," in *Handbook of Discourse Processes*, Mahwah, NJ, USA: Lawrence Earlbaum Associates, Inc., 2003, pp. 35–73.

36. United States Department of Agriculture, *Policies for USDA process verified program*, February 1, 2005 . Available: http://www.google.com/url?sa=t&rct=j&q=&esrc=s&source=web&cd=3&ved=0CDkQFjAC&url=http%3A%2F%2Fwww.ams.usda.gov%2FAMSv1.0%2Fgetfile%3FdDocName%3DSTELPRD3320447&ei=XpqDU4LIMIWTqAaZrIDQBg&usg=AFQjCNHfV51fJNa5IcVN-f71_1bY8Nj5Gg&sig2=D23syt05yeuEUKbjLsMLzg&bvm=bv.67720277,d.b2k

37. J.A. Scanga et al., "Development of computational models for the purpose of conducting individual livestock and premises traceback investigations utilizing National Animal Identification System-compliant data." *Journal of Animal Science*, vol. 85, pp. 503–511, 2007.

38. W. A. Price, "USDA's final rule on animal ID," *Farm-to-Consumer Legal Defense Fund*, January 18, 2013 . Available: http://www.farmtoconsumer.org/news_wp/?p=3862.

2

Children Communicating Food Safety/Teaching Technical Communication to Children: Opportunities Gleaned from the FIRST® LEGO® League 2011 Food Factor Challenge

Edward A. Malone and Havva Tezcan-Malone

FIRST LEGO® League (FLL) is an international robotics program for children in grades 4 through 8 (i.e., 9–14 year olds). Each year, FLL sponsors a themed challenge consisting of a robot game and a science project. The theme of the 2011 Food Factor Challenge was food safety. Based on our experiences with a team called the Global Dreamers in 2011, we argue that technical communicators can use FLL challenges to increase awareness and understanding of technical communication as a practice and a profession and that the 2011 Food Factor Challenge may offer a model for educating children about food safety if that model is integrated with instruction in technical communication.

2.1 Enhancing the Visibility and Recognition of Technical Communication

Technical communicators have long had an identity problem [1]. Most members of the public do not know what technical communication is or what a technical communicator does, sometimes even in the broadest terms. Even when terms such as *technical writing* and *professional communication* are used instead, misunderstandings abound. Technical communication is a practice, an academic discipline, and a profession. The practice

Communication Practices in Engineering, Manufacturing, and Research for Food and Water Safety, First Edition.
Edited by David Wright.
© 2015 The Institute of Electrical and Electronics Engineers, Inc. Published 2015 by John Wiley & Sons, Inc.

involves the communication of technical or scientific information through verbal and nonverbal means. The audience for such communication may be specialists or nonspecialists. Technical communicators have many different job titles, including technical editor, professional writer, medical illustrator, web designer, usability specialist, content developer or strategist, proposal manager, and information architect (to name just a few of the possibilities). Universities offer degrees in technical communication, but most students majoring in technical fields such as engineering or one of the sciences take a single technical communication course such as Technical Writing, Proposal Writing, Report Writing, or Introduction to Technical Communication.

Since the birth of the technical communication profession in the 1940s and 1950s, technical communicators have been struggling to build a mature profession and create an identity as recognizable in the public's mind as engineering [2]. To achieve this seemingly elusive goal, they have employed many means, ranging from creating professional associations to publishing journals and holding conferences to lobbying the government for changes in government job titles. They have also tried to bring visibility and recognition to technical communication by teaching technical communication in universities and (to a far lesser extent) in secondary schools. Almost no attention has been given to the teaching of technical communication in elementary and middle schools.

In this chapter, we argue that technical communicators can use FIRST® LEGO® League (FLL®)—an international robotics program—to increase awareness and understanding of technical communication as a practice and a profession—not only among the elementary and middle school students who participate in the program but also among the larger group of adults who serve as FLL judges, coaches, mentors, and spectators. Based on our experiences with a team called the Global Dreamers in the FLL 2011 Food Factor Challenge—which was organized around the theme of food safety—we came to realize that students who participate in one of the FLL annual challenges engage in as much technical communication as they do engineering and science and therefore need to be taught technical communication principles and skills in order to succeed in the competition. Moreover, it became clear to us that FLL itself does not fully realize the role that technical communication plays in its annual challenges. The term *technical communication* is not used in any of the FLL promotional and instructional materials about the annual challenge. Opportunities exist for technical communicators to support FLL teams as mentors and sponsors and increase awareness and understanding of technical communication as a practice and a profession.

In this chapter, we also suggest that the 2011 Food Factor Challenge—with some modifications—may offer a model for educating children about food safety in conjunction with technical communication. Not only did the challenge introduce FLL teams to the science and technologies of food safety, but it also underscored the crucial role of communication in making audiences aware of a contamination or spoilage risk and gaining the support of decision-makers for a proposed solution. Members of each team had to put themselves in the roles of food-safety scientists and engineers, conduct research into the production and/or transportation of a food item, and find a novel solution to a contamination or spoilage problem. More importantly for our purposes, they had to inform others about the problem and secure buy-in for their solution, and these tasks required extensive technical communication. Thus, the FLL challenge reinforced the children's learning of scientific and technological concepts by requiring them to communicate their research and ideas to diverse audiences, but it also gave them an opportunity to become better communicators—not merely by practicing their communication skills but also by learning how to communicate. This type of education has the potential to increase the effectiveness of food-safety measures in the future, especially if mentoring and instruction in technical communication are added to the FLL model.

> "The FLL challenge reinforced the children's learning of scientific and technological concepts by requiring them to communicate their research and ideas to diverse audiences."

In the sections that follow, we summarize relevant literature in technical communication, engineering, and food-safety education; we provide background information by explaining what FLL is, what the Food Factor Challenge was, and who the Global Dreamers were; we describe the technical communication activities involved in the project component of an FLL challenge and discuss the opportunities they present for technical communicators; and finally we argue that the Food Factor Challenge (the theme of the 2011 FLL competition) may serve as a model for an annual program for educating children in our country about food safety and introducing them to related careers.

2.2 Literature Review: Teaching Technical Communication, Engineering, and Food Safety to Children

For this project, the supporting literature falls into three groups: publications about the teaching of technical communication in elementary and secondary schools, publications about the use of FIRST LEGO League to teach engineering concepts and generate interest in engineering, and publications about food safety in relation to risk communication and the education of children. We believe that the publications in the first group are relevant to our discussion because they focus on the teaching of technical communication to precollege students, including students in elementary and middle school. The publications in the second group are relevant because they discuss the use of FIRST LEGO League in engineering education, both to teach foundational concepts and skills and to promote the profession of engineering. Our study is concerned with the use of FIRST LEGO League to teach technical communication concepts and skills and to promote the profession of technical communication. The publications in the third group are relevant because they establish the need for educating children about food safety and describe past and present educational programs in this area.

In the first group, several authors (e.g., References 3–5) have suggested strategies for integrating technical writing into high school curricula. Most of these suggestions hinge on effective collaboration between high schools and universities (e.g., teacher training programs). At various times in their histories, the National Council of Teachers of English (NCTE) and the Society for Technical Communication (STC) have reached out to secondary schools, providing resources and funding to teachers and information to students. For example, under the auspices of NCTE, Fearing and Allen [6] provided a resource for high school writing teachers who might want to create and implement an effective unit on technical writing. Several STC chapters have sponsored technical writing competitions for high school students and engaged in other outreach activities in secondary and even primary schools [7]. The goals of these outreach activities and collaborations range from giving back to the community and having fun [8] to recruiting future majors for a university program in technical communication [9], to improving the literacy levels of low-performing students [10, 11] with a shared goal of increasing awareness of technical communication as a practice and/or a profession.

In the literature we surveyed, only one publication discussed the teaching of technical communication in elementary and middle schools. Crawley [12] has described three activities that he uses to "spread the word about technical communication" when he visits local schools. The first activity, designed to meet the needs and interests of third graders, involved making ice cream cones. The students were given a definition of technical communication and shown some professional examples of technical writing; then they were asked to write a procedure for making an ice cream cone; and later they were given coupons for Dairy Queen®. The second activity, tailored to middle school students, explored why reading and writing are important and only incidentally included examples of technical communication. The third activity, geared toward high school students, involved writing a procedure for setting a mousetrap.

The literature in engineering education includes quite a few conference papers and journal articles about precollege robotics programs, some of which focus on FIRST LEGO League. In general, these papers recognize and value the interdisciplinary nature of FLL and the important role of well-informed mentors who can teach basic concepts and skills, particularly in relation to robotics and computer programming. They also recognize the value of FLL for grooming possible future engineers. Lau et al., for example, stated their opinion that a robotics program must go beyond introducing "engineering skills" and even exploring "scientific concepts and technological principles" to fostering "critical thinking, communications, and teamwork" [13, p. 12d4-26]. They favored "practical creative activities" over "traditional classroom methods," cooperation rather than competition between robots, and a nurturing attitude of "no [one] right answer" [14, p. 12d4-31]. Oppliger pointed to the potential usefulness of FLL as a way to promote the profession of engineering and "create a pool of future engineering students" [15, p. S4D-11]. Students at his university served as mentors to local FLL teams; he hoped the experience would improve their communication skills as well as help achieve other learning outcomes [16, p. S4D-13]. Howell et al. [17] acknowledged the interdisciplinary nature of FLL and suggested methods and sources of information to guide mentors in the teaching of engineering and programming concepts, but not concepts in other disciplines. Rusk et al. argued that robotics programs in general could attract more participants if they offered more "pathways" into robotics, and they suggested that such programs need to emphasize themes rather than challenges, pair engineering with art and even music, promote storytelling, and structure activities as "exhibitions rather than competitions" [18, p. 59]. None of these articles specifically mentioned the role of technical communication in FLL, but we feel they open the door for the observations and suggestions we make in this chapter. In the food-safety literature, there is ample recognition of the importance of effective risk communication in bridging the gaps between specialists and the public (e.g., [19, 20]). One important food-safety industry definition of risk communication is "the exchange of information and opinions concerning risk and risk-related factors among risk assessors, risk managers, consumers and other interested parties" [21]; (cf. Reference 22, p. 114). Note that consumers are involved in the exchange of information and opinions; in other words, they are involved in two-way communication (if not actual dialogue) rather than one-way communication. Risk communication, of course, is a form of technical communication (see, for example, [23,24]), and it was one of the types of technical communication that FLL teams engaged in during the Food Factor Challenge.

Many studies have underscored the need to educate middle schoolers about food risks and safe practices, ideally in a school setting (e.g., [25–27]). As one source stated, "Middle school is an ideal time to teach food safety for several reasons. Adolescents are in the process of setting lifelong behaviors; therefore, they are more likely to synthesize new food safety knowledge in a way that will lead to the development of lifelong behaviors" [28, p. 55]. Several national and international organizations actively support food-safety education for children, most notably the Partnership for Food Safety Education, which sponsors an educational program called "Fight BAC!®" for grades 4 through 8 [29]. Other food-safety education programs or initiatives for children include Food Safety: From Farm to Fork [30], Science and Our Food Supply: Investigating Food Safety from Farm to Table [31], and Hands On: Real World Lessons for Middle School Classrooms [32] as well as several computer-based games (e.g., [33, 34]). These programs and initiatives do not (to our knowledge) involve children communicating food safety in their communities.

2.3 Background: The League, the Challenge, and the Team

In 2011, a team called the Global Dreamers from Rolla, Missouri, competed in a regional tournament that was part of FLL's 2011 Food Factor Challenge. They built a robot, completed a science project, and competed with other teams at a tournament. Not only did the children and their mentors learn about food safety, but they also learned about technical communication, and we believe they would have learned even more if the role of technical communication had been recognized and emphasized in the handbook for FLL coaches and if technical communicators had been included as team mentors in the proverbial game plan. In this section, we will explain in more detail what FIRST LEGO League is, what the Food Factor Challenge was, and who the Global Dreamers were.

2.3.1 First Lego League

FIRST LEGO® League (FLL) is both a "robotics program" and an "organized sport" for children in grades 4 through 8 (i.e., 9–14-year-olds) in the United States and up to those aged 16 years old in most other countries [35]. FIRST, which stands for "For Inspiration and Recognition of Science and Technology," was started in 1989 by Dean Kamen, an inventor and entrepreneur [36]. In 1998, Kamen and Kjeld Kirk Kristiansen, owner of the LEGO Group, pooled their resources and expertise and started FLL [37].

Each year, FLL sponsors a challenge consisting of a robot game and a science project, both guided by a set of core values such as "We honor the spirit of friendly competition" and "What we discover is more important than what we win" [38]. Each FLL team may have as many as 10 children on it, and the children are assisted by an adult coach and sometimes one or more mentors, such as an engineer or a scientist [39]. "An FLL Mentor has a certain expertise a team may need" and "may contribute [his or her] expertise through instruction, guidance to the team, or serve as a resource on a one-time or multiple-time basis" [40]. The members of each team build a robot and compete in one or more tournaments; they also propose and communicate a solution to a scientific problem.

The core values and science project are just as important as the robot game. The tournament judges recognize excellence in each area with multiple awards, and the top award (i.e., the Champion's Award) goes to those who excel in all three areas [41]. FLL sums up the value of its annual challenge in this way:

> By designing our Challenges around such topics [as nanotechnology, climate, quality of life for the handicapped population, and transportation], participants are exposed to potential career paths within a chosen Challenge topic, in addition to solidifying the STEM principles that naturally come from participating in a robotics program. Team members also learn valuable life and employment skills which will benefit them no matter which career path they choose. [42]

In addition to FLL, FIRST also offers robotics programs for children in grades K-3 (Junior FIRST LEGO League®), grades 7–12 (FIRST Tech Challenge®), and grades 9–12 (FIRST Robotics Competition®), but FLL is unique in its combination of a robot game, a science project, and eight core values [43]. Both FLL and Jr. FLL® robots are made from LEGO bricks and other parts, but FLL teams also use LEGO MINDSTORMS® technology to design, build, and program their robots [44]. For our purposes, the most important distinction between FLL and its sister organizations is the science project, because it requires children to employ technical communication extensively.

About the 2011 FLL Challenge

Theme: Food safety

Science Project: To identify a food-safety problem and propose a solution

Robot Connection: Building and operating a robot to move miniature plastic food items around a game field—e.g., from field or stream to factory to store to kitchen (home base)

About Our Team (Global Dreamers)

The Members: Six boys and two girls between the ages of 11 and 14

Their Project (Problem and Solution): To track Juliet tomatoes back to the source of contamination or spoilage through the use of radio-frequency identification tags

Their Motto: Eat local, but think global—hence, the team's name "Global Dreamers"

Their Robot: Kismet, Jr., named after the famous robot at MIT

2.3.2 The Food Factor Challenge

Food safety was the theme of the 2011 FLL challenge. This theme organized the robotics game and informed the science project. The goal of both game and project was to ensure safe delivery of food for human consumption. The robotics game involved the performance of a series of missions such as "corn harvest," "pollution reversal," and "refrigerated ground transport" [45, pp. 12–13]. The instructions stated, "your robot's job is to put some common foods through just a few of the steps they go through in order to get into your belly, while either avoiding or dealing with contamination" [46, p. 12].

The science project focused on a contamination-related problem and solution and had three major steps:

1. *Selecting a Problem.* Each team member made a list of five foods found in his or her kitchen and investigated how each food item got to the kitchen and how contamination and spoilage were (or are being) prevented [47, p. 3]. Then the team members came together, compared their lists, and selected one food item to investigate collectively. They investigated that food item's journey from ground to table, noting all of the possible opportunities for contamination and spoilage as well as the actions taken to detect and avoid such problems. Finally, they chose one of the contamination or spoilage problems for their project [48, pp. 3–4].

2. *Devising a Solution.* After considering how professionals and others were solving the problem they had selected, the team had to devise a novel solution to the problem [49, pp. 4–5]. The solution might be as simple as inventing a more convenient way for workers to wash their hands before handling the food item or making an improvement to an existing storage container or safety protocol. The team had to be careful to propose a solution to a problem rather than a new detection method for identifying problems. The proposed solution became the focus of the team's technical communication activities.

3. *Communicating the Solution.* The team had to communicate their solution to multiple audiences. One group of audiences included people in the community (e.g., at a local company or in a government office). The FLL instructions suggested various ways to reach these audiences: "Give a talk. Create a website. Perform a skit. Make a comic book. Rap. Create a poster. Pass out flyers. Write

FIGURE 2.1. Members of the Global Dreamers. The children and their coach at the FLL Eastern Missouri Qualifier, Lindbergh High School, near St. Louis, November 12, 2011. All photos printed with permissions.

a poem, song, or story" [50, p. 4]. The other group of audiences included the judges and attendees at the FLL tournament. The team had to give a 5-minute presentation before a panel of judges who evaluated their presentation, research, solution, and so on against predetermined criteria. In the presentation, the team could employ "posters, slide shows, models, multimedia clips, your research materials—you are limited only by your team's creativity" [51, p. 5]. Before and after the presentation, some of these communication tools were on display for tournament attendees.

2.3.3   The Team: Global Dreamers

As stated above, an FLL team called the Global Dreamers—consisting of six boys and two girls between the ages of 11 and 14 (see Figure 2.1)—participated in the 2011 Food Factor Challenge. All eight children were students in either Rolla Middle School or Rolla Junior High School in Rolla, Missouri.[1] They were assisted by a coach (a part-time teacher who had a master's degree in physics), an assistant coach (a professional cook), a treasurer (a high school Latin teacher), and two mentors (a food expert and an engineer). All of the adults were parents of children on the team. The team chose to focus on the problem of locating the source and cause of contamination of Juliet tomatoes during transportation. They chose Juliet tomatoes because they are small and tasty and suitable for snacking by children. The team's proposed solution was to use radio-frequency identification (RFID) tags to capture information (such as origin and temperature fluctuations) as the tomatoes moved from farm to market. The judges at

---

[1] In Rolla, Missouri, middle school covers grades 5–7 and junior high covers grades 8 and 9.

FIGURE 2.2. Branding the team. The Global Dreamers designed a T-shirt (front and back) for their team and wore the T-shirts at the FLL Eastern Missouri Qualifier.

the FLL Eastern Missouri Qualifier determined that the team did not solve a food contamination problem, but rather proposed a means for detecting and responding to a contamination problem. Nevertheless, the team took second place out of 14 teams competing in the robot games and won a Judges Award as a rising star—that is, a team that the judges "expect great things from in the future" [52, p. 2].

2.4 Examples of Technical Communication Activities in FLL Projects

As mentioned previously, FLL teams are supposed to devise a solution to a scientific or technological problem (such as food contamination in the 2011 Food Factor Challenge) and communicate their solution to multiple audiences in their community and at tournaments. FLL encourages teams to employ diverse genres and media, such as a talk, skit, poster, website, slideshow, poem, song, and/or exhibit. Some of these forms (e.g., song, story) are usually classified as creative writing rather than technical communication, but their primary purpose in the challenge was to communicate technical or scientific information rather than just to entertain or delight.[2] Technical communication—that is, the communication of technical and scientific information—is not defined by the form it takes, but by its content and purpose.

Not only did the Global Dreamers use technical instructions and diagrams—authentic examples of technical communication—to build their robots and game field (see Figure 2.2), but they also employed technical communication in a variety of genres and media. For example, some members worked on a project website; other students designed posters and carried them to community events; still others wrote a skit involving a farmer, grocer, truck driver, and customer. Although the team practiced the skit in the library, they later decided not to use it at the FLL tournament because their presentation to the judges could not be longer than 5 minutes.

---

[2]On the use of poetry, stories (e.g., nonfiction narratives), and skits or plays in technical communication, see References 53–59, among other sources. For examples of plays about technical communication topics, see References 60, 61.

In this section, we have chosen to focus on four of the team's technical communication activities in order to highlight the prominent role that technical communication plays in FLL challenges:

- Branding (creating a name and logo)
- Conducting primary and secondary research
- Giving presentations and demonstrations
- Designing a document

2.4.1 Branding (Creating a Name and Logo)

Branding may refer to the entire effort of a company or an organization to communicate its identity (uniqueness, values, history, etc.), but it also refers to the creation of an effective name and logo. Branding is a technical communication activity when the identity involves science or technology.[3] FLL teams engage in branding by selecting their own names, designing their own logos, and communicating their identities to judges, spectators, other teams, and members of their communities. In the 2011 Food Factor Challenge, some teams had full-costumed mascots and gave out food gifts and team buttons at the tournament.

From one student's suggestion of "dreamers" and another's suggestion of "Think Global, Live Local" (a motto they had seen at the farmers' market), the team chose the name "Global Dreamers," in part because it captured the diverse backgrounds of the team members, who had first-generation family ties to other countries as well as the United States. Searching the Internet, a team member found an image of a robotic hand holding the world, and this image became part of the team's logo on a T-shirt (see Figure 2.3), because it seemed to suggest the present and future role of robotics and other technologies in protecting the environment and ensuring food safety. Green was chosen as the background color to suggest nature. Perhaps the design should have made some reference (an image, a color, a word) to Juliet tomatoes, but it did not.

2.4.2 Conducting Primary and Secondary Research

In the 2011 Food Factor Challenge, FLL teams were encouraged to interview professionals "who play a part in keeping food safe," such as farmers, fishers, chefs, butchers, truck drivers, agronomists, or botanists [69, pp. 1, 3]. Interviewing subject-matter experts (SMEs) is an important technical communication activity—a type of primary research

---

[3]The writing of technical sales literature has long been recognized as a specialization within technical communication. See, for example, References 62, 63. On branding as a technical communication activity, see References 64, 65. On the design of logos, see Reference 66 as well as other sources such as References 67, 68. The latter sources are sometimes used in desktop publishing courses in technical communication.

FIGURE 2.3. Following technical instructions. Members of the Global Dreamers use online instructions to assemble a dispenser for the replica game table.

that itself involves communication.[4] The Global Dreamers interviewed four farmers at a local farmers' market to learn about the risks of contamination: spoilage, loss, and predators to farm produce, such as eggs, tomatoes, and peppers (see Figure 2.4). The team also interviewed a WalMart® manager about the company's tracking of boxes of food, a General Mills® employee about the production of cereal, and a university professor about the technology of RFID tags. The children were like professional technical communicators in their reliance on SMEs for information to support and facilitate communication, but they were also like scientists, who supplement their own expert knowledge by doing field research.

To prepare for a project, a technical communicator also locates and uses secondary sources of information, such as legacy documents in company archives, published books and articles (whether in print or digital form), government sources, and university websites.[5] Although there is a distinction between research to support and facilitate communication of technical information and research to solve a science-related problem, in FLL projects this distinction is largely one of purpose, not type of research or research method, and in practice it is difficult to separate the two purposes. The students conduct secondary research much as a technical communicator who is not a SME does, and they use the research not only to select a problem and arrive at a novel solution but also to inform and develop their communications with audiences of nonspecialists

---

[4] Introductory textbooks in technical communication usually include a section about interviewing in a chapter or larger section about workplace research. For example, see Reference 70, pp. 140–142. Other sources on this topic include References 71–73.

[5] Most introductory textbooks in technical communication include a section about secondary research in technical communication. For example, see References 74, 75. See also Reference 76 for a longer (albeit dated) discussion.

FIGURE 2.4. Interviewing a subject-matter expert. Two members of the Global Dreamers interview a woman who raises chickens, harvests their eggs, and sells them at a farmers' market in Rolla, Missouri.

(e.g., a prospective sponsor in the community or a spectator at a tournament) as well as specialists or experts (e.g., some of the tournament judges).

In the 2011 Food Factor Challenge, FLL supplied a list of resources for secondary research. Most of them were websites about food safety: Kids World's Food Safety, Center for Food Safety, the Food Timeline, and FoodSafetyJobs.gov [77]. The remaining resources were magazine articles (all online, all from *Science News for Kids*), podcasts (all from the United States Department of Agriculture [USDA]), online videos (mostly from TED.com, but also from PBS, USDA, and Jamie Oliver), and print sources, (e.g., books about food handling, labeling, and production) [78]. The Global Dreamers used some of these resources, but they also located other (online) resources using Google® (see Figure 2.5).

The students in the 2011 Food Factor Challenge would have benefited from mentoring in search strategies, including note taking, paraphrasing, quoting, citing sources, fair use, and copyright laws. They were summarizing scientific information and presenting their solutions to the public through PowerPoint presentations, posters, and websites, but often they were using text that they had copied verbatim from online sources without acknowledgement. They also illustrated their presentations with images that may not have been in the public domain. The students were not always able to say where the images had come from, other than "the Internet."

2.4.3 Giving Presentations and Demonstrations

FLL requires teams to share their research and solution with audiences in the community, not just for practice, but to make a difference [79, p. 2]. Later, they must give a live

FIGURE 2.5. Conducting secondary research. Two members of the Global Dreamers are searching for information about RFID tags as a possible solution to the problem of controlling contamination of food items as they move from their points of origin to the consumer's table.

presentation at the tournaments in which they compete, and this presentation must adhere to certain criteria. For example, in the 2011 Food Factor Challenge, each team had to describe the research they conducted, the food item they chose to study, the contamination or spoilage problem they chose to address, their innovative solution, and the way they shared this solution with their community—all in 5 minutes or less [80]. The judges evaluated the presentation on effectiveness ("Message delivery and organization of the presentation"), creativity ("Imagination used to develop and deliver the presentation"), and sharing ("Degree to which the team shared their Project before the tournament with others who might benefit from the team's efforts") [81, p. 2].

Before the tournament, the Global Dreamers gave live project presentations and robot demonstrations throughout their community (see Figures 2.6 and 2.7). Their purpose was to promote awareness of FLL, attract sponsors, and solicit funds; their audiences included the board of directors of a local hospital, scientists and engineers at an electronics company, workers at a health food store, and the manager of a grocery store and a pizzeria, respectively. To accompany their presentation, they authored a series of PowerPoint® slides. The team was successful in obtaining the sponsors and contributions they needed; whether they were able to make a difference in their community is less clear. At the FLL Eastern Missouri Qualifier, the judges described their presentation as "very clear AND well organized" as well as "engaging and imaginative"; they also noted positively that the team had shared their proposed solution with "multiple individuals or groups" [82, p. 6].

FLL would like teams to become more confident and effective speakers as a result of their participation. One FLL document states that "Getting comfortable giving a 'live'

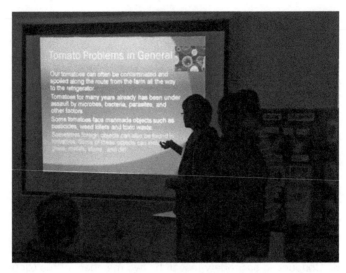

FIGURE 2.6. Delivering a presentation. The Global Dreamers created a series of PowerPoint slides and used them to explain their solution to a contamination problem.

FIGURE 2.7. Making a demonstration. On November 2, 2011, the Global Dreamers visited Brewer Science in Rolla, Missouri, and gave a science presentation as well as a demonstration of their robot, Kismet Jr.

presentation and sharing ideas effectively is a valuable life skill" [83, p. 1]. Presentations of technical and scientific information and demonstrations of technology are common forms of technical communication.[6] A mentor with training in technical communication would have much to offer FLL students as they plan, create, and deliver presentations and demonstrations. Such instruction would increase the likelihood that their public speaking experiences will be positive and successful.

2.4.4 Designing a Document

At tournaments, each team has a table displaying a cardboard exhibit that explains their problem and solution. Most of the exhibits are triptychs or tri-folds (i.e., consisting of three panels). Not only do competing teams and spectators move among the tables, looking at the different exhibits, but the judges also review the exhibits in their evaluations of the teams. Thus, the exhibits play an important role in the competition and must be designed to accommodate diverse audiences.

The exhibits at the 2011 Food Factor Challenge were interesting to us because they documented how other teams met the same FLL challenge. In general, the quality of the exhibits was much lower than the quality of exhibits at National History Day or similar regional and national events. FLL students are younger than the participants in those other events, and FLL students do not receive much guidance on how to design exhibits. The exhibits at the Eastern Missouri Qualifier tended to be text heavy, often with handwritten words. While the students created amazing robots, they could have been even more impressive if they had spent more time reading about the competition and planning their exhibits. One exhibit, for example, did not focus on the science project at all, but instead focused on the team's robot and was titled "Robot Game." Several exhibits did not have titles.

For their exhibit, the Global Dreamers chose to use the team's name as the main title and the phrase "Tracing the Juliet" as a subtitle (see Figure 2.8). They achieved a good balance of text and images in a symmetrical layout, but a major limitation of the design was the organization of the information: they did not employ a problem–solution pattern. In designing their exhibits, all of the teams could have benefited from a mentor who had some training and experience in document design and visual communication—two important areas of study in technical communication.[7]

2.5 The Food Factor Challenge as a Model of Food-Safety Education

The FLL challenge has evolved over the last decade into an effective educational program [100, 101], but in its present form it requires children to engage in fairly sophisticated

---

[6]On presentations in technical and scientific communication, see References 84–86. On demonstrations, see References 87–90. On the design and use of presentation slides, see References 91–96, among other sources.
[7]On document design and visual communication in technical communication, see References 97–99, among other sources.

FIGURE 2.8. Explaining their solution. The Global Dreamers created this tri-fold table exhibit and used it during the presentation to the tournament judges. Later, they displayed it in the Lindbergh High School gymnasium for FLL spectators.

technical communication without giving them the necessary preparation and guidance to do so. For this model to work even better, the children need mentoring and instruction in technical communication. A modified form of the Food Factor Challenge that includes such mentoring and instruction could serve the cause of food-safety education in the United States.

The Food Factor Challenge has much to offer the public as a model of food-safety education. It is an educational program that combines active learning strategies, a multidisciplinary approach, project-based activities, teamwork, collaboration, problem solving, and competition with an emphasis on cooperation and adherence to a code of conduct. It can be implemented locally under the supervision of parents and teachers, with sponsorship by schools and organizations. Ideally, it draws upon local scientists, engineers, and technical communicators as mentors, as well as people who work with food in various ways in the community, such as farmers, cooks, truck drivers, and grocery store managers. It is fun for the children and adults who participate.

An adapted form of the Food Factor Challenge—one that includes instruction and mentoring in technical communication—has the potential to solve several problems in food-safety education, such as how to eradicate dangerous behavior, how to improve the one-way flow of safety information, and how to attract more people into food-safety-related careers.

2.5.1 Fostering Food-Safety Habits in Children

One problem seems to be that it is more difficult to change bad behaviors in adults than foster good ones in children. People tend to form food-handling and -safety behaviors at an early age [102]. Ideally, we should be educating children about food safety before they enter elementary school, but the middle school years are particularly important because of that age group's "increasing responsibilities related to food preparation" [103, p. S51]. Providing food-safety education to people at an early age, as the Food Factor Challenge attempted to do, can be a means of cultivating positive, lifelong habits [104]. As Conley [105] points out, we in the United States do not want to have to teach the public about food safety during a crisis; we want to reinforce learning that has already taken place [105]. More importantly, though, we want to reduce the incidents of food-borne illnesses in the population at large, and one effective way may be through educational programs for children.

2.5.2 Promoting Dialogue Rather Than Monologue

Another problem in food-safety education seems to be that risk communication requires experts to engage in dialogue (rather than monologue) with members of a hazard-weary public [106]. As the Food and Drug Administration (FDA) noted, "risk communication with the public is a two-way street" [107]. Food-safety education is a form of risk communication as well as risk management; to be effective, it must be accomplished through dialogue [108]. But members of the public need to do more than just provide input and feedback to experts; they themselves must become effective science-based communicators of risk and agents of change in their homes and communities.

The Food Factor Challenge put children in the roles of risk communicators and some of their parents in the roles of coaches and mentors. Together, parents and children transformed science-based information into messages for community audiences, and they "tested" their messages on multiple audiences, fostering dialogue through question and answer sessions and critiques from audience members. They extracted information from SMEs through interviews. Children, in particular, may have the ability to break through the resistance of hazard-weary adults, who may be suspicious of government or distrustful of authority, and become "trusted and potentially politically neutral actors dispelling competing beliefs, convincing adults of new risks, and instilling more balanced views" [109, p. 41]. Children have been effective at communicating disaster and climate risk in their communities [110–113]. They might be equally effective communicators of food-related risk and safety measures in their homes and communities.

2.5.3 Generating Interest in Food-Safety Careers

Yet another problem seems to be the need to attract more qualified people to food-safety careers [114]. In general, K-12 students are unaware of career opportunities in food safety and do not become aware of these opportunities until college, if at all [115]. An educational program such as the Food Factor Challenge has the potential to increase awareness of and interest in food-safety careers as it teaches the relevant science in conjunction with technical communication. The Food Factor Challenge introduced participants to such careers as robotics engineer (i.e., one who might "design and maintain test instruments, food delivery systems, heating and cooling facilities, food processing facilities"), lawyer (who might "draft food safety regulations, review government requirements, prosecute food safety violators, advise corporations"), and agronomist (who might "oversee the health and safety of plants that are used for food") [116].

2.6 Conclusion

FLL offers many opportunities for professional technical communicators as well as university instructors and students of technical communication to engage in community service and promote their profession or academic discipline, working either individually or under the sponsorship of the STC, the IEEE Professional Communication Society, the Association of Teachers of Technical Writing, the Council of Programs in Technical and Scientific Communication (CPTSC), or one of the many university undergraduate and graduate programs in technical communication. Ultimately, though, the successful integration of technical communicators into FLL will depend on the willingness of FIRST to explicitly recognize technical communication as a profession and endorse technical communicators as prospective mentors of FLL teams in FLL promotional and technical documents, especially the handbook for coaches.

FLL would like its teams to become familiar with STEM careers: "Exposure to different areas of science and technology will also introduce team members to new career options they might never have known about. ... We want them to know about all the interesting fields of study that they may go into. ..." [117, p. 1]. One STEM-related field, of course, is technical communication, and the FLL activities in which the students engage are ideal for making them aware of and teaching them about careers in technical communication. Engineers are already using FLL and other robotics programs to identify, inspire, and track future engineers, and technical communicators should be doing the same.

The experience of FLL has shown us that students in middle school and junior high are not too young to learn technical communication concepts and develop technical communication skills. Working side by side with trained elementary and secondary teachers, technical communicators can help FLL teams communicate technical and scientific information effectively to diverse audiences in multiple genres and media. This instruction would complement their study of STEM subjects as well as the language arts.

The theme of food safety is not likely to be used again by FLL, and yet the Food Factor Challenge showed promise as a community-based program for educating children and their parents about food safety and introducing young people to the many career opportunities in food safety, including that of technical communicator. An FDA policy analyst who had participated as a judge in the Food Factor Challenge gave the following evaluation of the program:

> Increasing consumer awareness of food safety is one of the keys to preventing foodborne illness. The Food Factor Challenge accomplished so much more. It encouraged participants to learn about food safety; fostered creativity by challenging teams to come up with imaginative solutions to existing food safety problems; and inspired thousands—many of whom may become our future food scientists. [118]

Is it wise to abandon a potentially effective food-safety education program after a seemingly successful pilot run? An Agency such as the FDA, USDA, or CDC in cooperation with the Partnership for Food Safety Education and FIRST could adapt the program and run it again to see whether it has value as an annual (or at least recurring) event. The FDA, in particular, has experience with food-safety education for children. Not only does it offer online resources for "kids and teens" [119], but it engages in outreach to train "tweens" how to read nutrition facts labels [120] and continues to offer a summer training program for middle- and senior-level teachers and a free curriculum kit titled "Science and Our Food Supply" for classroom use (e.g., Reference 121). The FDA supported the Food Factor Challenge in 2011 [122].

The robot games are probably essential for motivating the children to participate. It seems unlikely that they would participate so enthusiastically if the program did not involve building a robot and competing with it at tournaments. But FIRST's linking of the robot games to the theme of food safety enhanced (rather than detracted from) the learning experience of the Food Factor Challenge. It gave the children an opportunity to consider the role of technology in food production and transportation. We suspect that the game field helped them to visualize and better comprehend some of the threats to food items on their varied journeys from farm or factory to table.

With informed mentoring in technical communication, the children who participate in this kind of program would have a better chance of meeting the expectations of live audiences and facing the challenges of the various communication situations in which they are placed, and they would develop their communication skills in tandem with their understanding of the relevant science. If they later chose careers in food safety, they would have a better understanding of the crucial role that communication plays in ensuring food safety and be better prepared to meet the communication challenges in their work.

Acknowledgments

The authors would like to thank the volume editor, David Wright, for the opportunity to write this chapter, and the series editor, Traci Nathans-Kelly, for her suggestions and contributions to this chapter. They would also like to thank the peer reviewers, Dan Voss and Susan Donohue, whose corrections and questions led to minor changes in the article. Finally, they would like to thank their colleagues: Patrick Allen, Jenny Bai, Brandon Clingenpeel, Sadhika Jagannathan, Lefatshe Lefatshe, Jesse Liu, Adem Malone, and Nikhil Wagher. This chapter could not have been written without these people.

References

1. D. Jones, "A question of identity," *Technical Communication*, vol. 42, no. 4, pp. 567–569, 1995.
2. E.A. Malone, "The first wave (1953–1961) of the professionalization movement in technical communication," *Technical Communication*, vol. 58, no. 4, pp. 285–306, 2011.

3. D. Gaskill, "History and theory of technical communication to guide collaboration between college and secondary school," in *Pathways to Diversity: 31st Annual Meeting of the Council of Programs in Technical and Scientific Communication*, Indiana, 2004, p. 71. Available: http://www.cptsc.org/pro/2004.pdf#page=78

4. K. Harley, "Perceptions and practices of technical communication by incoming college students and their secondary teachers," in *Pathways to Diversity: 31st Annual Meeting of the Council of Programs in Technical and Scientific Communication*, Indiana, 2004, p. 72. Available: http://www.cptsc.org/pro/2004.pdf#page=79

5. L. Zuidema, "Knowledge-Making in the development of high school technical communication programs: Who has a role?" in *Pathways to Diversity: 31st Annual Meeting of the Council of Programs in Technical and Scientific Communication*, Indiana, 2004, pp. 73–74. Available: http://www.cptsc.org/pro/2004.pdf#page=80

6. B.E. Fearing and J. Allen, *Teaching Technical Writing in the Secondary School*. Urbana, IL, USA: NCTE, 1984.

7. G. Lippincott and D. Voss, "Leveraging resources: How an STC chapter can support education in its community and professional development for its members," *Technical Communication*, vol. 48, no. 4, pp. 449–464, 2001.

8. C.R. Crawley, "From ice cream to mousetraps: Explaining technical communication to k-12 students," in *A Global Odyssey: Proceedings of STC's 48th Annual Conference*, Illinois, 2001, pp. 23–26.

9. J.D. Ford, "Strategies for attracting high school students to technical communication programs," in *Pathways to Diversity: 31st Annual Meeting of the Council of Programs in Technical and Scientific Communication*, Indiana, 2004, p. 88. Available: http://www.cptsc.org/pro/2004.pdf

10. T.R. Grill, "Documentation as problem solving for literacy outreach programs," *Lawrence Livermore National Laboratory, USA, Livermore, CA*, Jul 9, 2004 [Online]. Available: https://e-reports-ext.llnl.gov/pdf/309320.pdf

11. T.R. Grill, "Building science-relevant literacy with technical writing in high school," *IEEE Transactions on Professional Communication*, vol. 49, no. 4, pp. 346–353, 2006.

12. C.R. Crawley, "From ice cream to mousetraps: Explaining technical communication to k-12 students," in *A Global Odyssey: Proceedings of STC's 48th Annual Conference*, Illinois, 2001, pp. 23–26.

13. K.W. Lau, H.K. Tan, B.T. Erwin, and P. Petrovič, "Creative learning in school with LEGO® programmable robotics products," in *29th Annual Frontiers in Education Conference*, San Juan, Puerto Rico, 1999, vol. 2., pp. 12D4/26–12D4/31.

14. K.W. Lau, H.K. Tan, B.T. Erwin, and P. Petrovič, "Creative learning in school with LEGO® programmable robotics products," in *29th Annual Frontiers in Education Conference*, San Juan, Puerto Rico, 1999, vol. 2., pp. 12D4/26–12D4/31.

15. D. Oppliger, "Using FIRST LEGO league to enhance engineering education and to increase the pool of future engineering students (work in progress)," in *32nd Annual Frontiers in Education Conference*, Boston, 2002, vol. 3, pp. S4D/11–S4D/15.

16. D. Oppliger, "Using FIRST LEGO league to enhance engineering education and to increase the pool of future engineering students (work in progress)," in *32nd Annual Frontiers in Education Conference*, Boston, 2002, vol. 3, pp. S4D/11–S4D/15.

17. W.L. Howell, E.J. McCaffrey, and R.R. Murphy, "University mentoring for FIRST LEGO league," in *33rd Annual Frontiers in Education Conference*, Boston, 2003, vol. 2, pp. F3A/11–F3A/16.

18. N. Rusk, M. Resnick, R. Berg, and M. Pezalla-Granlund, "New pathways into robotics: Strategies for broadening participation," *Journal of Science Education and Technology*, vol. 17, no. 1, pp. 59–69, 2008.

19. D.A. Powell and W. Leiss, Eds., *Mad Cows and Mothers Milk: The Perils of Poor Risk Communication*. Quebec, Canada: McGill-Queens University Press, 1997.

20. T.L. Sellnow, R.R. Ulmer, M.W. Seeger, and R. Littlefield, *Effective Risk Communication: A Message-Centered Approach*. New York, NY, USA: Springer, 2009.

21. World Health Organization, "The application of risk communication to food standards and safety matters: Report of a joint FAO/WHO expert consultation," Rome, Italy: FAO, 1999. Available: http://www.google.com/search?tbo=p&tbm=bks&q=isbn:9251042608

22. Codex Alimentarius Commission, "Procedural manual." Rome, Italy: World Health Organization, 2013. Available: ftp://ftp.fao.org/codex/Publications/ProcManuals/Manual_21e.pdf

23. B.A. Sauer, *The Rhetoric of Risk: Technical Documentation in Hazardous Environments*. Mahway, NJ, USA: Lawrence Erlbaum Associates, 2003.

24. R.E. Lundgren and A.H. McMakin, *Risk Communication: A Handbook for Communicating Environmental, Safety, and Health Risks*, 5th ed. Hoboken, NJ, USA: John Wiley & Sons, 2013.

25. I. Haapala and C. Probart, "Food safety knowledge, perceptions, and behaviors among middle school students," *Journal of Nutrition Education and Behavior*, vol. 36, no. 2, pp. 71–76, 2004.

26. C. Byrd-Bredbenner, V. Quick, and K. Corda, et al., "Middle schoolers: Food safety cognition and behaviors [poster abstract]," *Journal of Nutrition Education and Behavior*, vol. 44, no. 4, pp. S51, 2012.

27. M. Barclay, K. Greathouse, M. Swisher, S. Tellefson, L. Cale, and B.A. Koukol, "Food safety knowledge, practices, and educational needs of students in grades 3 to 10," *Journal of Child Nutrition and Management*, vol. 27, no. 1, pp. 72–75, 2003. Available: http://docs.schoolnutrition.org/newsroom/jcnm/03spring/barclay/

28. J. Richards, G. Skolits, J. Burney, A. Pedigo, and F.A. Draughon, "Validation of an interdisciplinary food safety curriculum targeted at middle school students and correlated to state educational standards," *Journal of Food Science Education*, vol. 7, no. 3, pp. 54–61, 2008.

29. Partnership for Food Safety Education, "Your game plan for food safety: Teacher's activity and experiment guide: A Fight BAC! Food safety education program for 4th, 5th and 6th grade classrooms." Available: http://fightbac.org/storage/documents/curriculum/fight%20bac%20curriculum%20book.pdf

30. California Foundation for Agriculture in the Classroom, "Agriculture in the classroom online: Food safety—from farm to fork" [Online], 2012. Available: http://www.cfaitc.org/foodsafety/

31. Food and Drug Administration and the National Science Teachers Association, "Curriculum kit: Science and our food supply" [Online], 2013. Available: http://www.teachfoodscience.com/curriculum.asp

32. J. Richards, C. Pratt, G.J. Skolits, and J. Burney, "Developing and evaluating the impact of an extension-based train-the-trainer model for effectively disseminating food safety education to middle school students," *Journal of Extension*, vol. 50, no. 4, 2012. Available: http://www.joe.org/joe/2012august/a6.php

33. R.A. Lynch, M.D. Steen, T.J. Pritchard, P.R. Buzzell, and S.J. Pintauro, "Delivering food safety education to middle school students using a web-based, interactive, multimedia, computer program," *Journal of Food Science Education*, vol. 7, no. 2, pp. 35–42, 2008.

34. V. Quick, K. Corda, and C. Byrd-Bredbenner, "Middle schoolers' food safety cognitions and intended behaviors: A food safety computer game intervention," *Journal of the Academy of Nutrition and Dietetics*, vol. 112, no. 9, pp. A–5, 2012.

35. FIRST LEGO League, "Support our mission," 2013. Available: www.firstLEGOleague.org/mission/support

36. FIRST, "About us: Vision and mission," 2013. Available: www.usfirst.org/aboutus/vision

37. FIRST LEGO League, "Our founders," 2013. Available: firstlegoleague.org/mission/founders

38. FIRST LEGO League, "Core values," 2013. Available: firstLEGOleague.org/mission/corevalues

39. FIRST LEGO League, "Support our mission," 2013. Available: www.firstLEGOleague.org/mission/support

40. FIRST, "Mentor and coach roles," 2013. Available: www.usfirst.org/community/volunteers/mentor-coach-role

41. FIRST LEGO League, "Awards descriptions," 2014. Available: www.firstlegoleague.org/sites/default/files/Official_Event_Info/Awards%20Descriptions2014.pdf

42. FIRST LEGO League, "Challenge: Overview and history," 2013. Available: www.firstLEGOleague.org/challenge/thechallenge

43. FIRST, "FIRST progression of programs," 2013. Available: http://www.usfirst.org/roboticsprograms

44. FIRST, "FIRST progression of programs," 2013. Available: http://www.usfirst.org/roboticsprograms

45. FIRST LEGO League, "Food factor challenge," 2011. Available: http://firstLEGOleague.org/sites/default/files/Challenge/FoodFactor/FLL2011_Complete_Challenge.pdf

46. FIRST LEGO League, "Food factor challenge," 2011. Available: http://firstLEGOleague.org/sites/default/files/Challenge/FoodFactor/FLL2011_Complete_Challenge.pdf

47. FIRST LEGO League, "Food factor challenge," 2011. Available: http://firstLEGOleague.org/sites/default/files/Challenge/FoodFactor/FLL2011_Complete_Challenge.pdf

48. FIRST LEGO League, "Food factor challenge," 2011. Available: http://firstLEGOleague.org/sites/default/files/Challenge/FoodFactor/FLL2011_Complete_Challenge.pdf

49. FIRST LEGO League, "Food factor challenge," 2011. Available: http://firstLEGOleague.org/sites/default/files/Challenge/FoodFactor/FLL2011_Complete_Challenge.pdf

50. FIRST LEGO League, "Food factor challenge," 2011. Available: http://firstLEGOleague.org/sites/default/files/Challenge/FoodFactor/FLL2011_Complete_Challenge.pdf

51. FIRST LEGO League, "Food factor challenge," 2011. Available: http://firstLEGOleague.org/sites/default/files/Challenge/FoodFactor/FLL2011_Complete_Challenge.pdf

52. FIRST LEGO League, "Awards descriptions," 2014. Available: http://www.firstlegoleague.org/sites/default/files/Official_Event_Info/Awards%20Descriptions2014.pdf

53. B.F. Barton and M.S. Barton, "Narration in technical communication," *Journal of Business and Technical Communication*, vol. 2, no. 1, pp. 36–48, 1988.

54. J.J. Connor, "Poetry at work: Historical examples of technical communication in verse," *Journal of Technical Writing and Communication*, vol. 18, no. 1, pp. 11–22, 1988.

55. S. Glassman, "The technical writer as playwright," *Technical Writing Teacher*, vol. 14, no. 1, pp. 118–119, 1987.

56. J. McCoy and H. Roedel, "Drama in the classroom: Putting life in technical writing," *Technical Writing Teacher*, vol. 12, no. 1, pp. 11–17, 1985.

57. W. Quesenbery and K. Brooks, *Storytelling for User Experience: Crafting Stories for Better Design*. N.p.: Rosenfeld Media, 2010.

58. J.M. Perkins and N. Blyler, *Narrative and Professional Communication*. Stamford, CT, USA: Ablex Publishing Company, 2000.

59. R. Rutter, "Poetry, imagination, and technical writing," *College English*, vol. 47, no. 7, pp. 698–712, 1985.

60. J.M. Lufkin, "The slide talk: A tutorial drama in one act," *IEEE Transactions on Engineering, Writing and Speech*, vol. 11, no. 1, pp. 7–14, 1968.

61. J.M. Lufkin, "The fatal abstract: A tutorial farce in one act," *IEEE Transactions on Engineering, Writing and Speech*, vol. 14, no. 1, pp. 3–9, 1971.

62. H.W. Smith, Jr., "An analytical approach to the development of technical sales literature," *Journal of Technical Writing and Communication*, vol. 4, no. 3, pp. 207–215, 1974.

63. F.T. Van Veen, "Technical sales literature," in *Handbook of Technical Writing Practices*, vol. 1, S. Jordan et al., Eds. New York, NY, USA: Wiley-Interscience, 1971, pp. 581–614.

64. S. Harner and T. Zimmerman, "Internal and external branding," in *Technical Marketing Communication*. Boston, MA, USA: Longman, 2002, pp. 67–86.

65. C. Pettis, *TechnoBrands: How to Create & Use "Brand Identity" to Market, Advertise & Sell Technology Products*. New York, NY, USA: Amacom, 1995.

66. R.C. Parker, *Looking Good in Print*, 6th ed. Scottsdale, AZ, USA: Paraglyph Press, 2006.

67. B. Hunt, "Some great web logo designs for inspiration," 2006. Available: http://www .webdesignfromscratch.com/web-design/great-web-logos-for-creative-inspiration/

68. L. Silver, *Logo Design that Works: Secrets for Successful Logo Design*. Gloucester, MA, USA: Rockport Publishers, 2001.

69. FIRST LEGO League, "Ask a professional," 2011. Available: http://firstLEGOleague .org/sites/default/files/Challenge/FoodFactor/FLL2011%20_Project_Ask_a_Pro.pdf

70. M. Markel, *Technical Communication*, 10th ed., New York, NY, USA: Bedford/St. Martin's, 2012.

71. M. Flammia, "The challenge of getting technical experts to talk: Why interviewing skills are crucial to the technical communication curriculum," *IEEE Transactions on Professional Communication*, vol. 36, no. 3, pp. 124–129, 1993.

72. M.F. Lee and B. Mehlenbacher, "Technical writer/subject-matter expert interaction: The writer's perspective, the organizational challenge," *Technical Communication*, vol. 47, no. 4, pp. 544–552, 2000.

73. E.E. McDowell, *Interviewing Practices for Technical Writers*. Amityville, NY, USA: Baywood Publishing, 1991.

74. H. Graves and R. Graves, "Researching technical subjects," in *Strategic Guide to Technical Communication*, 2nd ed. Ontario, Canada: Broadview Press, 2012, pp. 73–94.

75. L.J. Gurak and J.M. Lannon, "Strategies for researching on the internet," in *Strategies for Technical Communication in the Workplace*. Boston, MA, USA: Longman, 2012.

76. L.R. Porter and W. Coggin, *Research Strategies in Technical Communication*. New York, NY, USA: John Wiley & Sons, 1995.

77. FIRST LEGO League, "Resources," 2011. Available: http://firstLEGOleague.org/sites/ default/files/Challenge/FoodFactor/FLL2011_Project_Resources.pdf

78. FIRST LEGO League, "Resources," 2011. Available: http://firstLEGOleague.org/sites/default/files/Challenge/FoodFactor/FLL2011_Project_Resources.pdf

79. FIRST LEGO League, "Why a project in a robotics competition?" 2011. Available: http://firstLEGOleague.org/sites/default/files/Challenge/Why_a_Project_in_a_Robotics_Competition.pdf

80. FIRST LEGO League, "Food factor challenge," 2011. Available: http://firstLEGOleague.org/sites/default/files/Challenge/FoodFactor/FLL2011_Complete_Challenge.pdf

81. FIRST LEGO League, "Judges' rubrics," 2014. Available: http://www.firstlegoleague.org/sites/default/files/Official_Event_Info/Combined%20Rubrics2014.pdf

82. FIRST LEGO League, *First Lego League 2011 Food Factor Project Judging Pre-Tournament Preparation Pack, Smaller Qualifying Tournament Edition* [Online], 2011. Available: http://4hset.unl.edu/4hdrupal/sites/default/files/competitions/2011_FLL/PRSmallPack.pdf

83. FIRST LEGO League, "Why a project in a robotics competition?" 2011. Available: http://firstLEGOleague.org/sites/default/files/Challenge/Why_a_Project_in_a_Robotics_Competition.pdf

84. M. Alley, *The Craft of Scientific Presentations*. New York, NY, USA: Springer-Verlag, 2003.

85. L. Gurak, *Oral Presentations for Technical Communicators*. Boston, MA, USA: Allyn and Bacon, 2000.

86. H.J. Scheiber and P.J. Hager, "Oral communication in business and industry: Results of a survey on scientific, technical, and managerial presentations," *Journal of Technical Writing and Communication*, vol. 24, no. 2, pp. 161–180, 1994.

87. R.J. Brockmann, "Using emulation to teach nonverbal technical communication in the twenty-first century," in *Proceedings of the 43rd Annual Int. Technical Communication Conference of the Society for Technical Communication*, Seattle, 1996, pp. 8–13.

88. R.J. Brockmann, "19th Century oral technical communication: NCR's silver dollar demonstration," in *From Millwrights to Shipwrights to the Twenty-First Century: Explorations in a History of Technical Communication in the United States*. Cresskill, NJ, USA: Hampton Press, 1998, pp. 99–117.

89. J. Care and A. Bohlig, "The dash to demo," in *Mastering Technical Sales: The Sales Engineer's Handbook*. Boston, MA, USA: Artech, 2002, pp. 85–98.

90. R.L. Sullivan and J.L. Wircenski, "Presenting a technical demonstration," in *Technical Presentation Workbook: Winning Strategies for Effective Public Speaking*, 2nd ed. New York, NY, USA: Asme, 2002, pp. 158–171.

91. M. Alley and K.A. Neeley, "Rethinking the design of presentation slides: A case for sentence headlines and visual evidence," *Technical Communication*, vol. 52, no. 4, pp. 417–426, 2005.

92. M. Alley, M. Schreiber, K. Ramsdell, and J. Muffo, "How the design of headlines in presentation slides affects audience retention," *Technical Communication*, vol. 53, no. 2, pp. 225–234, 2006.

93. J. Doumont, "The cognitive style of PowerPoint: Slides are not all evil author," *Technical Communication*, vol. 52, no. 1, pp. 64–70, 2005.

94. J. Mackiewicz, "Audience perceptions of fonts in projected PowerPoint text slides," *Technical Communication*, vol. 54, no. 3, pp. 295–307, 2007.

95. J. Mackiewicz, "Comparing PowerPoint experts' and university students' opinions about powerpoint presentations," *Journal of Technical Writing and Communication*, vol. 38, no. 2, pp. 149–165, 2008.

96. E.R. Tufte, *The Cognitive Style of PowerPoint: Pitching Out Corrupts Within*, 2nd ed. Cheshire, CT, USA: Graphic Press, 2006.

97. E.R. Brumberger and K.M. Northcut, *Designing Texts: Teaching Visual Communication.* Amityville, NY, USA: Baywood Publishing, 2013.

98. K.A. Schriver, *Dynamics in Document Design: Creating Texts for Readers.* New York, NY, USA: John Wiley & Sons, 1997.

99. C. Kostelnick and D.D. Roberts, *Designing Visual Language: Strategies for Professional Communicators.* Boston, MA, USA: Allyn and Bacon, 1998.

100. Brandeis University for FIRST, "Executive summary: Evaluation of the FIRST LEGO League 'Climate Connections' Season (2008–09)," September 2009. Available: http://www.usfirst.org/sites/default/files/uploadedFiles/Who/Impact/Brandeis_Studies/2009_FLL_Brandeis_University_Evaluation_Executive_Summary.pdf

101. A. Melchior, T. Cutter, and F. Cohen, "Evaluation of the FIRST LEGO league underserved initiative," July 2004. Available: http://www.usfirst.org/uploadedFiles/Who/Impact/Brandeis_Studies/05FLL_Underserved_Full_Report.pdf

102. M. Barclay, K. Greathouse, M. Swisher, S. Tellefson, L. Cale, and B.A. Koukol, "Food safety knowledge, practices, and educational needs of students in grades 3 to 10," *Journal of Child Nutrition and Management*, vol. 27, no. 1, pp. 72–75, 2003. Available: http://docs.schoolnutrition.org/newsroom/jcnm/03spring/barclay/

103. C. Byrd-Bredbenner, V. Quick, and K. Corda, "Middle schoolers: Food safety cognition and behaviors [poster abstract]," *Journal of Nutrition Education and Behavior*, vol. 44, no. 4, p. S51, 2012.

104. I. Haapala and C. Probart, "Food safety knowledge, perceptions, and behaviors among middle school students," *Journal of Nutrition Education and Behavior*, vol. 36, no. 2, pp. 71–76, 2004.

105. S. Conley, "Science, not scares: Communicating food safety risks to 'hazard-weary' consumers," *Food Safety and Inspection Service, United States Department of Agriculture*, 1998. Available: http://www.fsis.usda.gov/OA/speeches/1998/sc_iamfes.htm

106. S. Conley, "Science, not scares: Communicating food safety risks to 'hazard-weary' consumers," *Food Safety and Inspection Service, United States Department of Agriculture*, 1998. Available: http://www.fsis.usda.gov/OA/speeches/1998/sc_iamfes.htm

107. Food and Drug Administration, "Strategic plan for risk communication," 2009. Available: http://www.fda.gov/AboutFDA/ReportsManualsForms/Reports/ucm183673.htm

108. World Health Organization, "The application of risk communication to food standards and safety matters: Report of a joint FAO/WHO expert consultation." Rome, Italy: FAO, 1999. Available: http://www.google.com/search?tbo=p&tbm=bks&q=isbn:9251042608

109. T. Mitchell, T. Tanner, and K. Haynes, "Children as agents of change for disaster risk reduction: Lessons from El Salvador and the Philippines," *Brighton, UK: Institute of Developmental Studies*, 2009. Available: http://www.undpcc.org/undpcc/files/docs/publications/CCC_Working%20Paper_Final1_Screen.pdf. Accessed on June 24, 2015.

110. M.A. Hamad, "The child-to-adult method in mine-risk education," *Journal of Mine Action*, vol. 11, no. 1, 2007. Available: http://www.jmu.edu/cisr/journal/11.1/notes/aziz/aziz.shtml

111. Institute of Development Studies, "Children communicating climate and disaster risks," *IDS in Focus: Policy Briefing* 13.3, November 2009. Available: http://www.ids.ac.uk/files/dmfile/IF13.3.pdf

112. T. Mitchell, K. Haynes, N. Hall, W. Choong, and K. Oven, "The role of children and youth in communicating disaster risk," *Children, Youth and Environments*, vol. 18, no. 1, pp. 254–279, 2008. Available: http://www.jstor.org/stable/10.7721/chilyoutenvi.18.1.0254

113. T. Mitchell, T. Tanner, and K. Haynes, "Children as agents of change for disaster risk reduction: Lessons from El Salvador and the Philippines," *Brighton, UK: Institute of Developmental Studies*, 2009. Available: http://www.undpcc.org/undpcc/files/docs/publications/CCC_Working%20Paper_Final1_Screen.pdf. Accessed on June 24, 2015.

114. J. Garden-Robinson and K. Beauchamp, "Attracting the next generation of food safety professionals through marketing and education in high schools poster abstract," *Journal of Nutrition Education and Behavior*, vol. 44, no. 4, pp. S52, 2012.

115. M. Wiedmann and S.K. Warchocki, "Aim 2—Develop and deliver k-12 food safety activities," *Cornell University*, 2011. Available: https://confluence.cornell.edu/display/FOODSAFETY/Aim+2+-+Develop+and+deliver+K-12+food+safety+activities. Accessed on June 24, 2015.

116. FIRST LEGO League, "*Ask a professional*", 2011. Available: http://firstLEGOleague.org/sites/default/files/Challenge/FoodFactor/FLL2011%20_Project_Ask_a_Pro.pdf. Accessed on June 24, 2015.

117. FIRST LEGO League, "*Why a project in a robotics competition?*" 2011. Available: http://firstLEGOleague.org/sites/default/files/Challenge/Why_a_Project_in_a_Robotics_Competition.pdf. Accessed on June 24, 2015.

118. M. Naum, "The next generation of food safety innovators," *FDA Voice*, 22 June 2012. Available: http://blogs.fda.gov/fdavoice/index.php/2012/06/the-next-generation-of-food-safety-innovators/

119. Food and Drug Administration, "Food safety & nutrition information for kids and teens," 2013. Available: http://www.fda.gov/Food/ResourcesForYou/Consumers/ucm2006971.htm

120. Food and Drug Administration, "Nutrition facts label: Read the label youth outreach campaign," 2014. Available: http://www.fda.gov/Food/IngredientsPackagingLabeling/LabelingNutrition/ucm281746.htm

121. Food and Drug Administration and the National Science Teachers Association, "Curriculum kit: Science and our food supply," 2013. Available: http://www.teachfoodscience.com/curriculum.asp

122. M. Naum, "The next generation of food safety innovators," *FDA Voice*, 22 June 2012. Available: http://blogs.fda.gov/fdavoice/index.php/2012/06/the-next-generation-of-food-safety-innovators/

3

The Role of Public (Mis)perceptions in the Acceptance of New Food Technologies: Implications for Food Nanotechnology Applications

Mary L. Nucci and William K. Hallman

3.1 Accepting New Foods: Consumers, Technology, and Media

Many consumers firmly believe, as Brillat-Savarin wrote in 1826, "Tell me what you eat, and I will tell you what you are." Through our consumption of food, we literally internalize the positive or negative credence qualities we perceive related to foods. Because they are ingested, perceived food risks represent hazards that are both physically and symbolically incorporated into one's corporal and spiritual being [1].

Indeed, an extensive literature on the anthropology of food, eating, and culture illustrates that food can be used to define social identity [2], illuminate political-economic value creation, understand the social construction of memory [3–7], [8,9], provide explanations for human behavior [8,9], define the relationship between cultural and biological evolution [10–14], [15,16], demonstrate gender, personal, and national identity [17–22], [8], and identify economic and political changes [23–28]. Food can act as a social change agent [29–31] and is powerfully linked to social life [32, 33], war, and ethnic conflict [34–39].

Communication Practices in Engineering, Manufacturing, and Research for Food and Water Safety, First Edition.
Edited by David Wright.
© 2015 The Institute of Electrical and Electronics Engineers, Inc. Published 2015 by John Wiley & Sons, Inc.

Feasting, fasting, and the ritual preparation and consumption of certain foods and taboos or restrictions regarding the touching or eating of other foods all play crucial roles in religious and cultural practices, and identities [40–47]. Moreover, people often use their food choices to represent and communicate *who* they are as individuals, or their roles in society or to express their political or ideological beliefs [48–54]. Giving or sharing food with others is symbolically, psychologically, and emotionally linked with love, nurturing, and intimacy, and is considered crucial to creating and maintaining bonds between people [55–57].

> Food carries with it distinct religious, symbolic, and cultural meanings that set it apart from other concerns [58].

Because foods are so very different from most other products, consumers' concerns about food risks must be considered separately from the perceptions of the hazards present in other domains [59–64]. Consumers and experts also perceive food risks differently [65, 66]. Indeed, acceptance of food technologies is strongly tied to a variety of factors that are extrinsic to the product, including cultural, social, and moral variables related both to the product and to the prospective consumer of the product [60, 67, 68]. Cardello [67] noted, "In the case of novel foods or foods that have been processed by novel or emerging technologies, concerns about the nature of the food and/or the nature of the processing technologies that have been used to treat the food become paramount considerations."

Other studies show that a consumer's basic orientations to food and the kinds of food he or she already eats are important predictors of their receptivity to new food technologies [69, 70]. Consistent with this, a recent study of New Zealanders' intentions to buy beef and lamb products which were altered to have 20% less cholesterol using nanotechnology found that while the perceived benefits of the product were important, issues of self-identification, perceived social support, and perceived barriers to purchase were also important in predicting who would be willing to buy the products presented [71]. In Chile, a decreased willingness to purchase nanotechnology foods was associated with a higher level of food neophobia, less satisfaction with life, and less satisfaction with food-related lifestyle behaviors (i.e., shopping, cooking, production information) [70].

3.1.1 Food Technology Acceptance

In consequence, new food technologies often encounter problems with successfully entering the consumer market due to perceived public perceptions and concerns. For example, in the case of the proposed use of food irradiation, consumers did not oppose either its objective (to make food safer) or the entities involved (the killing of pathogens) or the qualities of the resulting food products. Instead, beliefs about the radiation technology involved [72, 73] appeared to be a powerful factor for lack of consumer acceptance despite strong support from the scientific community, including the American Medical Association, Scientific Committee of the European Union, the World Health Organization, the Centers for Disease Control and Prevention, the United States Public Health

Service, and the Institute of Food Technologists, all of which deemed it safe and effective [73, 74].

Similarly, with the use of carbon monoxide packaging to prevent oxidation of packaged beef products, both the technology (i.e., carbon monoxide) and the perceived objective (i.e., to extend the saleable shelf life of beef by prolonging the appearance of freshness) were seen as problematic, despite being recognized as safe by the US Food and Drug Administration (FDA) and promoted by the meat industry as a safe and effective method of extending the acceptable shelf life of meats by preventing "premature browning" [75]. Complaints by consumer groups describe the use of carbon monoxide packaging as unacceptable [76–78], arguing that it is both deceptive and risky because it can mask visible signs of spoilage [78]. Moreover, some consumers, understanding carbon monoxide as a deadly gas, may find the idea of having meat exposed to a gas they understand to be poisonous as either dangerous or disgusting [79, 80].

Consumers also expressed concern with the use of recombinant bovine somatotropin (rBST) for increased milk production [81, 82] when it was approved in 1993 by the Food and Drug Administration. This was despite its being recognized as safe for human consumption [83, 84]. Consumers' belief in the wholesomeness of milk [85] and its promotion as a primary food for infants and children led to heightened perceptions of risk for milk consumption [86]. Milk produced using rBST was seen as both artificial and unfamiliar [87]; combined with a lack of comprehensible information from the regulatory agencies [88] and a greater sense of risk for synthetic versus natural products [85], consumers were reluctant to purchase and consume milk from rBST-treated cows. In a survey of consumers, 80% said they were worried about future health effects [87], indicating that they assigned a greater risk to rBST milk than the experts [81]. Today, consumers and consumer groups continue to express concern about its use in milk and milk-based products [89].

Importantly, the public's acceptance of new technologies, especially those with wide-ranging applications, is not homogenous, but rather can diverge sharply depending on use. When the public perceives clear benefits from an application, such as biotechnology-based treatments for Parkinson's disease or diabetes, acceptance is higher than for applications where benefits are not as clearly apparent, such as with genetically modified (GM) food [90].

Indeed, support for genetically modified products is sharply divided across a "red-green" divide, with stronger support for red (medical applications) GM products as opposed to green (agricultural applications) GM products [91].

In regards to food biotechnology, the research points to a complex of factors that impact acceptance. Much of the research on GM foods has indicated that there is a "hierarchy of approval" [92–94] regarding its characteristics, uses, and benefits that affects consumers' acceptance. GM ingredients within some categories of food products were more acceptable than others. Consumers were more willing to eat processed or "junk" foods containing GM [70] while they would be unwilling to purchase GM milk or baby food [95]. Aside from health, environmental, and production-related benefits [96], issues of trust related to the food industry, government and media, ethics, democracy, uncertainty, and morality also play a role in support for GM foods [97–102].

Consumers also differed by country in terms of acceptance of genetic modification [103]. Europeans tend to be more negative toward genetic modification compared to North Americans [104]. It is speculated that this is due to issues in valence of communications in the European as compared to US press [105] and that US consumers have had a positive experience with the larger number of GM products available on the market in the United States [106].

In the case of genetically modified food, the origin of the DNA with which plants and animals are modified mattered to people, with consumers preferring that both the origin and target were similar in nature. They were least open to DNA crossing species and were less willing to purchase GM foods where novel DNA did not become part of the final product [94]. Overall, consumers found the application of biotechnology to plant products more acceptable than its application to meats and animal products [93–95, 107, 108], likely due to moral and ethical issues related to "tampering with nature" or "playing God" [98, 109].

3.1.2   The Role of the Media in Public Perceptions of Food Technologies

Communication of new developments in food processing plays a critical role in public perceptions and adoption of new food technologies. The media are the main source through which the public is apprised of scientific findings, controversies, and policy discussions and "influence[s] the perceptions of a public increasingly sensitive to the social and ethical implications of science and inclined to question the credibility of scientists and technical institutions" [110, p. 1603]. Most adults, once they have left formal education, learn about science from newspapers, television, and the Internet [111]. Thus, the presentation of science by the media—the content, scope, and style—can affect the public perception of science and public support for legislation and policy [112].

> For journalists, fresh and dramatic research that allows them to entertain as well as inform is a motivating factor in newsworthiness. The news values of journalists privilege controversy: "crisis, the individual event over the past or future, and conflicts" [113, p. 43].

Due to cultural norms and differences of opinion of their respective roles, journalists and scientists process and present scientific information in very different ways. They diverge in their definitions of what is newsworthy about science, their respective styles of communication, and their vision of the media's role [110]. Journalists often have little to no science training, making them ill-equipped to analyze science claims or competing knowledge [114]. Science stories are generally presented as polarized and two sided [10, 115], rather than nuanced or complex [116], leading even science journalists to complain that their own reporting is unnecessarily streamlined and misleading [117,118]. News coverage often removes the caveats and limitations typical to science papers [10, 119–121] and presents science as more certain than it is [122].

Articles must appeal to the readers and the editor. Routine science is less exciting than scandalous science. Nisbet, Brossard, and Kroepsch noted that "issues that receive the greatest media attention are those that are most easily dramatized or narratized" [113, p. 43).

Unfortunately, as the work of science straddles expert and public spheres, these differences in communication styles have led to conflicts in the public acceptance of scientific developments. These cultural differences are further compounded by the growing realization by scientists that they are unable to control the media's or the public's interpretations of science. Frames influence how individuals interpret information, and in consequence, public opinion instead is shaped by the frames selected and presented in the news media [123–126]. The more positive response to medical applications of genetic modification may be linked to more positive media coverage over a 12-year period than media coverage of the agricultural applications of genetic modification[124]. Readers of science stories that emphasized the social or political aspects of science felt more risk than readers of stories that emphasized the scientific or technical aspects [127] or provided risk comparisons to support decision-making [128].

Once a frame is established, it can be very difficult to shift the issue to another perspective, and the ability to control the dominant frame can have long-term consequences. For example, it has been shown that journalists who tend to favor larger institutional sources over activist or special interest groups include organizational spokespersons more frequently in articles about science [129, 130]. These journalistic practices can lead to a social perspective that certain voices or issues matter while others do not [131]. Since its inclusion in media coverage, GM news coverage is more likely to use an industry, commercial, or government source progress frame in stories in mainstream media [132–136]. This pro-biotechnology frame rarely included other points of view [137] except for events that triggered negative news [133, 138]. It is argued that this reliance on institutional sources created a sense that on the whole the US public was accepting of GM foods and that opposition was confined to a small number of extremists [139]. Activist groups are still primarily only successful in promoting specific issue frames [140], in responding to non-routine events [141] or in placing single pieces, than in influencing the framing of an issue [142–148]. Even when included in coverage, these alternate voices have a limited role in the development of science or technology-related issues [131].

3.1.3 The Case of GM and the Media

The genetic modification of food is a key example of how information dissemination has impacted public perceptions of the technology. Prior to the 1990s GM food received little attention in US media [113, 149, 150]. The majority of any news coverage about genetic modification tended to emphasize benefits and downplay ethical, legal, social impact or regulatory issues [105, 151]. Coverage became less positive in the 1990s due to events such as the cloning of Dolly the sheep and the outbreak of mad cow disease [133, 152–154], and resulted in a reframing of the discussions to include ethics, risks, and accountability [133]. But on the whole, GM food was a nonissue in the mainstream media throughout that decade.

Likely as a function of the dependence on news for scientific information [114], US citizens had little knowledge of GM food and were ambivalent toward its application [94, 95]. To a great degree this continues today, in spite of the ubiquity of GM food in processed food sold at grocery stores [93].

Nisbet and Lewenstein [133] commented in their analysis of the media coverage of biotechnology between 1970 and 1999 that specific events can have a profound impact on publications patterns and emphases. They wrote that "Changing events can not only shift the balance of source influence, but they can also introduce new frames to a debate that may mobilize or allow access to interests previously not included in the media and policy agenda-building process" (ibid, 8).

Of the events that prompted a shift in public perceptions of genetic modification, the publication of research on the toxic effect of pollen from *Bacillis thuringiensis*-modified corn on the Monarch caterpillar is considered to be a watershed moment due to its influence on the tone and volume of coverage in US media [133, 154–157]. In May 1999, John Losey, Linda Rayor, and Maureen Carter [158] published a scientific correspondence in the journal *Nature*, showing that pollen from this genetically modified corn killed Monarch butterfly larvae. In keeping with journal policies, *Nature* issued a press release on the findings. Cornell University, where the research was performed, also issued a press release. Although the research was preliminary in nature, and had not been replicated, the Union of Concerned Scientists (UCS), Greenpeace (both anti-genetic modification activist groups), and the Biotechnology Industry Organization (BIO, a pro-genetic modification industry organization), all also issued press releases on the published correspondence. Each text had a particular frame: the BIO press release was pro-biotech, the Losey et al. communication was cautionary, and both the Greenpeace and UCS press releases were anti-biotech in tone. The Pew Initiative on Food and Biotechnology [156, p. 3] noted that the Losey et al. letter sparked a "worldwide controversy" that consequently generated national and international news coverage that transformed the Monarch butterfly into a symbol of the anti- genetic modification movement.

An examination of the newspaper coverage of this research over the year after its publication showed that the majority of stories appeared in the first month of coverage in the 19 papers examined, with more than half appearing on the first page of the first section of the newspapers examined. This is in line with observations of a media spike associated with this event [150, 154, 159]. Over the entire year examined, the newspaper stories tended to use language consistent with the original Losey et al. [158] article or Greenpeace press release, and not of the BIO press release, the industry organization. This is in contrast to previous studies of newspaper coverage of biotechnology, which have shown that the story line is dominated by the use of industrial and university frames, which tend to focus on economic considerations and potential benefits [133, 138].

In communicating information about a new food technology, it is very important to consider how science information is translated from the scientist to general public, and what is or is not lost in that translation. In this particular study, the frame of the Losey et al. [158] correspondence was as likely to be picked up by the press throughout the media coverage of this issue as that of the activist frame and was more commonly used than the industry frame. This was in contradiction to the expected industry frame predominance, as has been shown by other researchers [133]. For biotechnology in general, a shift in frame emphasis was beginning to develop after 1997, with ethics and public accountability frames starting to become more prominent [159]. The results of this study support this conclusion, as the industry frame was used less often than the Losey frame or the activist frame. However, the shift in frame selection for this incident was as much influenced by the trend toward ethics and public accountability as it was by the fact that the incident in question—the contamination of human food with unapproved food products—had potential widespread effects. Dorothy Nelkin [110] has shown that medical information that affects a large number of people receives more media coverage. Though the contamination of food is not a medical issue, the concept of the contamination of food constitutes a potential health crisis, which was further compounded by the possibility of allergic responses to the genetically engineered component in the tainted food.

Additionally, the Monarch butterfly is well known throughout most of the United States: they are found in all of the 48 lower states, citizen science programs like Monarch Watch (www.monarchwatch.org) promote awareness, and the Monarch is used extensively to teach schoolchildren about biology [160], so the possible death of the Monarch from genetically engineered pollen would affect a large number of people. Gamson and Modigliani [161] noted that the relative importance of media discourse depends on how readily available meaning generation experiences are in people's everyday lives. The story of the death of the Monarch could be construed as a universal controversy for US public. What was a preliminary study with an appeal for more research as a science correspondence in *Nature* became an emotional story of the loss of the Monarch butterfly.

3.2 Nanotechnology: Unseen, Unknown

As a developing technology within the agriculture and food sectors, the adoption of nanoscale technologies is expected to have impacts on farmers and food that will "exceed that of farm mechanization or of the Green Revolution," changing every step of the food chain from "soil to supper" [162, p. 1]. New seed technologies, veterinary drug and vaccine delivery systems, the nano-encapsulation of pesticides and herbicides ("nanocides"), improved fertilizers and soil amendments, and "Little Brother Technology" surveillance systems consisting of nanoparticle sensors designed to monitor, record, and communicate the presence or absence of moisture, soil nutrients, pathogens, and threatening chemicals are either in development or already in distribution [163].

Nanotechnology is also creating a revolution in food packaging. Researchers are developing new edible coatings and laminate films to serve as barriers to moisture, lipids, and gasses, extending the shelf life of foods [164–166]. Such nanolaminate coatings or films could also serve to improve the texture of foods or as a way to carry functional agents designed to enhance the flavor, color, nutrition, health benefits, and safety of food products. Others are working on specialized bioanalytical nanotech sensors designed to be incorporated in food packaging with the ability to indicate the presence of particular nutrients or signs of spoilage, tampering, or contamination by heavy metals and other harmful chemicals, pathogens, particulates, or allergens [167, 168].

While some nanotech products are already on the market, much of the rhetoric related to nanotechnology is associated with its future promise [169]. Indeed, the idea of nanotechnology, like its predecessor agricultural biotechnology, is associated with the promise to end hunger, have little environmental impact, expand the production of food, and at the same time, meet consumer desires for food products [170].

3.2.1 Nanotechnology in the Media

As discussed earlier, the media are a critical source of information on new technologies, and for food nanotechnology, will likely play a key role in shaping public perceptions and acceptance [171]. The way that the media frames the discussions around nanotechnology is important, because through it, the media not only sets the boundaries of debates around scientific issues [135, 172, 173] but also plays an important role in establishing the perceptions of risk and benefit related to new technologies [174].

Nanotechnology has been the focus of media attention since at least 2000, when the government formed the National Nanotechnology Initiative, a research and development coalition of 20 federal department and agencies. The level of coverage has tended to increase over time [175, 176] but with a year-to-year variation in total coverage [177]. Early newspaper coverage of nanotechnology between 1990 and 2004 was generally positive and was primarily framed as scientific discovery/projects, progress or business [178–180]. Articles that were focused on ethical, legal, or social implications (ELSI) of the technology or business were not as frequent [175, 180]. The tone of coverage of nanotechnology in US news through 2005 remained generally positive, focusing on progress and potential economic benefits [175, 181–183]. Even in articles exploring risk, the number of positive assessments was about equal with the number of negative assessments [179].

Consumer-related themes such as quality, safety, and packaging increased in 2007 and 2009 [171], and concerns related potential toxicity, environmental damage, regulations, and other uncertainties relevant to nanoscale materials began to appear more frequently after 2005 [175, 181, 182, 184–187]. Regardless, throughout the years, the media frame for nanotechnology has been optimism related to the benefits mixed with uncertainties related to both benefits and risks [181].

3.2.2 Public Perceptions of Nanotechnology

Despite this coverage, the existing literature on public perceptions of nanotechnology suggests that most US citizens currently know little about it, and many lack familiarity even with the term "nanotechnology" [188–191]. In 2004, the majority of US citizens (80%) indicated that they had heard little to nothing about nanotechnology [189], as a result, survey participants had minimal success in correctly answering factual questions about the technology. In a 2005 survey, 54% reported that they knew nothing about nanotechnology, 17% felt they knew something about nanotechnology, and 26% said they knew a little [192]. In studies performed in 2005 and again in 2007, there was no increase in the awareness of nanotechnology [193, 194]. Only 6% said they had heard a lot about nanotechnology; 92% had heard some to nothing at all [194]. Through

2009, knowledge of nanotechnology remained unchanged [195, 196] and in a Harris poll conducted in 2012, US Citizens still reported to have heard only a little about nanotechnology [197].

Regardless, most studies have indicated that despite knowing little about nanotechnology, the public tends to have a positive or neutral perception of the technology, [178, 190–200]. There is some evidence that the balance of benefits and risks is shifting to more risk concerns [196]; even without having a clear mental model of the technology, opinions about it appear well developed [199]. When asked about specific applications of nanotechnology, US consumers were willing to use certain products even with health or safety risks; for example, nanotechnology was perceived to be less risky than herbicides, chemical disinfectants, handguns, and food preservatives [201].

However, it has been shown, as similar to genetic modification technologies, that the specific benefit and perceived use of nanotechnology is critical to acceptance, with medical applications ranking highest (31%) among consumers followed by better consumer products at 27%, general progress (12%), environmental protection (8%), safer and better food (6%), energy, economy, electronics (4%), and benefits to soldiers and security [192] at 3% overall. When specific applications of food-related nanotechnology were presented to Swiss consumers, barcodes for food security, ultraviolet (UV) protection packaging, and stronger packaging film were ranked highest for benefits and lowest for risks, while nanotech-modified foods, livestock food infused with protein to promote animal health, and nutritional supplement capsules were ranked lowest for benefits and highest for risks [202]. In the same study repeated in Mexico, barcodes for guaranteed food security, health-promoting green tea, bread with nutritional value, UV-protection packaging, and cancer-preventing supplements were found highest for benefits and lowest for risks, while individually modifiable foods, bacteria detection spray, oxygen-blocking plastic bottle, stronger packaging film, and antibacterial food containers were found lowest for benefits and highest for risks [203].

Unfortunately, measures of perceptions and intentions have been shown to be unreliable predictors of behavior [204, 205]: what people say they will do, and what they actually do are often very different. Even when perceived benefits are high, such as with nanotechnology packaging, acceptance is low [201], indicating that benefit alone does not guarantee acceptance. This is similar to findings from a study of consumer acceptance of GM foods, where even clearly communicated benefits still resulted in low willingness to consume [206].

3.2.3 Perceptions and Acceptance of Nanotechnology

At issue for acceptance of nanotech-enabled products is that the lack of knowledge of nanotechnology may require consumers to draw upon analogies to past technologies, such as asbestos, dioxin, Agent Orange, or nuclear power that confounds acceptance [192] or personal perspectives on naturalness, religion, science, technology and nature [189], [195], [204], [207–209], or the food industry [208, 211, 212]. Critically, because of consumers' lack of knowledge about nanotechnology, decisions about its acceptance or rejection beg the question "will the fate of nanotechnology be determined by rumor and supposition, as some believe has been the case for GMO? Or will public opinion be based on objective science and engineering findings?" [201, p.153].

> Perceptions of nanotechnology are critical to its acceptance, because the success of the early promise of nanotechnology is as much dependent on consumers' openness to and acceptance of nanotech products as it is on the ability to create them.

Public apprehensions about the applications of new technologies to foods may be exacerbated by the mental models and concerns they have about the basic materials or methods employed to achieve the resulting benefits. For example, the lack of consumer acceptance of food irradiation technology is often blamed, in part, on the public's inability to separate the concept of irradiation from that of radiation. Thus, the negative affective responses many have toward radiation, and specifically toward the thought of food potentially contaminated by radiation, led many consumers to reject the idea of irradiated foods [213].

Experience with the introduction of new functional foods and nutraceuticals also suggests that consumers may be most receptive to product enhancements that are congruent with the way they think about and use the original product. For example, sales data suggest that consumers are more receptive to the addition of cholesterol-lowering plant sterols to a product such as orange juice (which is often consumed for its health benefits) than in a product such as margarine [214].

Thus, the preconceived mental models consumers hold about the nature of the materials that may come into contact with their food may play a strong role in their perceived acceptability. To further complicate matters, communications about nanotechnology are confounded by the complexity of the science and the broad ranges of applications [215]. By definition nanotechnology operates using invisible materials with properties very different from the same macro-level materials with which consumers are more familiar. As such, the real and perceived risks and uncertainties associated with the technology may present substantial reasons for consumers to reject nanotech products.

Examining Consumer's Acceptance of Nanotechnology Food In 2010, we conducted a series of interviews to assess consumers' perceptions and acceptance/rejection of food-related nanotechnology applications, endeavoring to mimic the path toward decision making about an unknown field: exposure to the subject, information gain about the subject, and access to the subject (through descriptions of hypothetical nanotechnology-enabled food products). Qualitative interviews were used to evaluate latent knowledge about nanotechnology in general and food nanotechnology specifically before and after being given the opportunity to read the National Nanotechnology Initiative (NNI) brochure, *Nanotechnology: Big Things from a Tiny World* [216]. Launched in 2000, the NNI is composed of 25 federal agencies with a range of research and regulatory roles and responsibilities in nanotechnology and is funded through the federal budget (National Nanotechnology Initiative, undated a). This NNI-designed brochure is described as "A colorful, compelling brochure that explains nanotechnology and its potential in a format that appeals to general audiences" [216] and is intended to provide a general overview of basic concepts of nanotechnology from the federally funded NNI project.

Consistent with previous studies of consumer's knowledge of nanotechnology [192–217], we found that very few of the participants reported knowing anything about nanotechnology prior to reading the NNI brochure, noting comments such as they "Honestly can't think of anything," "Don't know technology," "It doesn't ring a bell," "I've heard of it but I really don't know anything. I'm sure someone will go 'nano nano,'" or "This would be something I couldn't figure out and understand. I'm not technologically savvy."

When asked what came to mind when they heard the word "nanotechnology," participants mentioned technologies such as computers, iPods, chips, junk drives, or lasers; medical therapies such as miniature cameras for in vitro examination and monitoring or repairing human cells; and popular culture, including *Mork and Mindy* ("nanu nanu": the character Mork used the phrase "nanu nanu" to say hello), James Bond, and *Stargate*. Only 3 out of 31 total participants linked nanotechnology to food and food preparation. One participant felt that nanotechnology was used for "Instant fractions of a second, computer-generated devices to help make decisions about food quality or food analysis," while another commented that it had something to do with food composition. Referencing their mention of *Mork and Mindy*, the third participant commented that "I think of a planet like ours with fields of vegetables really, and—something flat like a satellite flying over and just, like, getting images of these elements that, whatever, vegetables or what have you." Notably, later in the interview two of the three participants acknowledged that they had been influenced by the flier used to recruit participants, which specifically mentioned new food technologies.

Many knew that the term "nanotechnology" was somehow linked to concepts of size ("Because I just think of something little when you say 'nano.' I think my kids have something that's 'nano' and its small."; "Well, I don't know much about it, but 'nano' usually means something small."; "The smaller it is, the better."; "Very small but intelligent."), often associating it with the benefits of miniaturization. Participants commented, "I guess it's compressing more and more and smaller and smaller and smaller"; "It can do all the same stuff, but it comes in a much smaller package, like almost like a fifth of the size or less"; "Nanotechnology is just making things smaller"; and "Assuming I'm correct about the miniaturization, then I would say that it's—you know, an evolution of an existing technology that has, you know, somehow become miniaturized."

However, no participant was able to describe the key attributes of nanotechnology identified in most authoritative definitions, that is, the ability to engineer, understand, and manipulate matter at the nanoscale, and the capability to generate and make use of the novel properties of materials at the molecular level [218, 219]. When asked specifically about familiar products that currently used nanotechnology, very few were able to provide any examples; the majority ($N = 21$; 68%) could not name *any* products. Categories of products thought to incorporate nanotechnology included clothing (shirts), electronic (microwaves, mp3 players, computer chips, cell phones, video games, hard drives, computers, iPods), medical (miniature cameras, pharmaceuticals), or industrial applications (robotics, carbon fiber tubes) were mentioned by several participants.

Participants had more difficulty when asked about the connection between nanotechnology and food, often struggling to come up with a linkage. Several mentioned processed, engineered, or efficient food as an outcome of using nanotechnology, while others felt that the use of nanotechnology was associated with healthy foods, making foods better, downsizing meals, nutraceuticals, or low calorie foods. Focusing on the technology aspect, a number of participants felt that nanotechnology would be connected to food through new or novel cooking technologies, microwaves, or cooking equipment; or improvements in agricultural production through increased yield, hybridization of plants or animals, food safety/food quality monitoring, prevention of illness in animals, or replacement of pesticides. Quick meals, futuristic foods, unhealthy/processed foods, food production, or food safety were each mentioned by a single participant. One participant commented, "I really doubt there would be too much nanotechnology in the food we actually eat. That I think would be pushing the boundaries a little. ... If people don't know what's inside their food—you know, that's technically, I think, breaking the law."

After reviewing the NNI brochure, slightly less than half of the participants used terms about size (such as "small," "minimize," "shrink," or "miniaturization"), instead of mentioning green concepts related to cleaning up the environment or nature or concepts such as cost-effectiveness, improving products, greater efficiency, better solutions, better materials, higher standards, stronger, lighter, or safer. One participant noted that there was "not much about food" in the brochure. In fact, there were only three references to food/water applications of nanotechnology in the brochure: "Cosmetics and food producers are 'nano-sizing' some ingredients, claiming that improves their effectiveness" [217, p. 4], "which has been shown to neutralize bacteria, including *E.coli*, in water" (ibid, 6), and "low cost technology for cleaning arsenic from drinking water" (ibid, 8).

It was not surprising, therefore, that there was no significant increase in the number of participants who mentioned associations between food and nanotechnology after reading the NNI brochure. Following on the concepts of efficiency and cost-effectiveness prevalent in the brochure, participants noted that nanotechnology was linked to food through time and size concepts such as growing food faster or easier, growing bigger fruits and vegetables, or increasing yield, concepts of purification, or detection or prevention of contamination. Other participants linked nanotechnology to food through food engineering (taste/color/quality and creating new foods), production of healthier food, new and improved cooking technologies, and packaging.

Significantly, we found that before reading the brochure, nearly all of the associations that participants mentioned with regard to nanotechnology were either positive or neutral. Only one individual commented in the negative, saying, "Because I don't want to think about eating technology. I mean, I think it should be a choice if you're going to put something in you." After reading the brochure, only about a third raised unprompted concerns about the potential health impacts of nanotechnology; primarily concerns about side effects, long-term impacts, and overall safety. Moreover, the percentage of unprompted statements expressing concern was essentially unchanged after tasting the food products. However, when prompted by specific questions as to whether the participant had any health, religious, ethical/moral, environmental, or other concerns about nanotechnology and food, 84% of participants noted issues with the application of the technology to food products.

3.3 Discussing New Food Technologies

The successful adoption of new food technologies is not guaranteed, regardless of scientific support, economic value, enhanced benefits, and minimum risks. This can be seen in the history of rejected technologies and, more recently, in the public skepticism and hostility that accompanied the introduction of biotechnology [220]. The literature is replete with calls by both scientists and public commentators to heed the lessons learned from the launch of biotechnology and not to repeat the same mistakes with nanotechnology and future food technologies [221–228].

The public now demands participation in scientific and technical decisions in a move from an uncritical acceptance of new science and technology to solve social and historical problems to one of social scrutiny linked to the perceptions of risks and benefits of development [229–232]. This shift in uncritical acceptance of science developments can be understood through growing social and consumer activism. Failing to take into consideration the social and cultural factors that drive perceptions and support for science and refusing to acknowledge the voice of the public has given many citizens good reasons to doubt the trustworthiness of science and technology [229, 233].

Critical for science is that decisions can no longer be based solely on objective data-based calculations [234]. Strong associations regarding the perceptions of differing technologies center the debates around developing technologies on issues of science, technology, and society [190]. Decision-making needs to integrate the knowledge, skills, economic interests, and moral values of the public into scientific governance [235].

Indeed, deep political controversies, ethical debates, and trade wars have arisen over the real and perceived health, environmental, ethical, religious, and cultural consequences of the successful development and dissemination of new technologies and their products. Consequently, the "Field of Dreams Philosophy," that is, "If you build it, he will come" [236], does not necessarily apply to the products of emerging technologies [224]. In part, this is because consumers do not evaluate technologies or products based solely on the promise of their benefits, but rather assess the trade-offs between both risks and benefits within a context of an understanding (or misunderstanding) of how the technology works, what it is intended to accomplish, and what adoption of the technology may mean for both culture and society. While the Internet has made scientific information much more widely available to people around the world, simply making scientific information available is not enough to increase either knowledge or acceptance of science and technology. Moreover, while scientific literacy remains low worldwide, increased knowledge about science has been repeatedly shown to be a poor predictor of the acceptance of controversial technologies [74, 237–239].

Instead, studies show that public perception and acceptance or rejection of new science and technology is based in large part on individual and societal concerns about issues that go beyond the science itself and on answers to questions that cannot be directly addressed by science. These trans-scientific issues [240] include social, political, and cultural factors that are influenced by overall worldviews, views about nature, risk tolerance, trust in institutions, religious beliefs and religiosity, personal values, and cultural background, among many others [133, 191, 201, 210, 211, 237, 241–246].

These shared cultural norms are important in the formation of perceptions that drive the acceptance or rejection of technologies [247].

Nanotechnology is an "invisible" technology. The general public knows very little about it, and its presence cannot be seen or touched. Unlike other technologies, nanotechnology has no clearly defined predecessor. The technology is not just about making things smaller, as it is making things smaller on a new scale and the changes that scale effects on the compound [248]. Therefore, the public will be forced to make a decision to accept or reject the technology based on some complex calculus of personal beliefs and perceptions about science and technology. Similar to the public's response to genetic modification, the response to nanotechnology will likely be one where differences in age, education, gender [93], ethnicity, religion, and trust in scientists, corporations, and government [93, 249], all become important factors in decision making.

More specifically for food nanotechnology, the consumer's orientation to food and the kinds of food they eat [69], as well as environmental attitudes [217] and social values [250] may be important predictors of receptivity to new food technologies. A study of New Zealanders' intentions to buy beef and lamb products made using nanotechnology found that while the perceived benefits of the product were important, issues of self-identification, perceived social support, and perceived barriers to purchase were also important in predicting who would be willing to buy the products presented [71]. More work will need to be done to consider the role that these factors play in the acceptance of nanotechnology food applications.

In our study, we found that participants were willing to spend a considerable time talking about nanotechnology as part of the interview and learning about it through reading the NNI brochure. Yet it was the specific application of the nanotechnology, its perceived associated benefits, and the hedonic and sensory qualities of the resulting food products that appeared to have the greatest influence on consumer responses to food nanotechnology than their apparent understanding, approval, or disapproval of the technology itself. Most studies have indicated that despite knowing little about nanotechnology, the public currently has a positive or neutral perception of the technology, with perceived benefits outweighing potential risks [178, 199, 201]. It may be that the novelty of the technology combined with an inability to grasp it in human scale or connect it to previous constructs makes it still somewhat protected from negative perceptions. Many US consumers have indicated a willingness to use nanotech products even though they may have health or safety risks, finding that nanotechnology was less risky (and of equal or higher benefit) than food preservatives, chemical disinfectants, chemical fertilizers, DDT, and pesticides [202].

Indeed, for most of the interview participants, the concept of nanotechnology did not appear to carry any inherent specific negative valence or stigma, while the potential benefits of the products described were clearly evaluated along a continuum. This suggests that in judging the application of nanotechnology to food products, the participants were more likely to appraise the benefits of the foods (which they had little trouble grasping) than to evaluate the technology (which they did not appear to clearly understand).

Regardless, the participants expressed a strong desire to know more about nanotechnology in order to make a decision about actually purchasing or ingesting any nano food product or food product packaged in nano-enhanced materials, *especially* about its potential short-term and long-term risks. They made it clear that while they were willing to try nanotechnology food products, for many of them future purchasing decisions would depend on information regarding the long-term effects of eating foods made with nanotechnology: "public perceptions of nanotechnology are not as simple as previously assumed—risks and benefits are both enmeshed in a complex decision-making calculus" [206, p. 154]. Skepticism about particular applications of food nanotechnology may reflect that this calculus of mental constructs—which includes a complete lack of knowledge of the technology and misperceptions generated by science fiction books, television, and movies—and known product benefits simply does not outweigh the unknown risks of the technology. Therefore, consistent with cautious consumer reactions to the introduction of other novel food technologies, efforts to market nanotech foods based solely on the benefits of the technology are unlikely to lead to consumer acceptance. It behooves all those involved with disseminating information about nanotechnology, and specifically food nanotechnology, to take into account the public's lack of mental constructs in making decisions about acceptance or rejection of this developing technology.

Acknowledgments

Research discussed here was funded by Grant 2008-01415 to the Food Policy Institute, Rutgers, the State University of New Jersey, from the Cooperative State Research, Education, and Extension Service of the United States Department of Agriculture (USDA): *Food Nanotechnology: Understanding the Parameters of Consumer Acceptance*, Dr. W. Hallman, principal investigator; and Grant 2002-52100-11203 from the US Department of Agriculture (USDA), under the Initiative for the Future of Agricultural Food Systems (IFAFS): *Evaluating Consumer Acceptance of Food Biotechnology in the United States*, Dr. William K. Hallman, principal investigator. The opinions expressed are those of the authors and do not necessarily reflect official positions or policies of the USDA, the New Jersey Agricultural Experiment Station, or of the Food Policy Institute, Rutgers, the State University of New Jersey.

References

1. A. Angyal, "Disgust and related aversions," *The Journal of Abnormal and Social Psychology*, vol. 36, no. 3, pp. 393–412, 1941.
2. L. Wilken and A.L. Knudsen, "Milk, myth and magic: The social construction of identities, banalities and trivialities in everyday Europe," *KONTUR*, vol. 17, pp. 33–41, 2008.
3. A.H. Akram-Lodhi and C. Kay, *Peasants and Globalization: Political Economy, Agrarian Transformation and Development*. New York: Routledge, 2012.
4. M. Dietler and B. Hayden, *Feasts: Archaeological and Ethnographic Perspectives on Food, Politics, and Power*. Tuscaloosa, AL: University of Alabama Press, 2010.

5. P. McMichael, "A food regime genealogy," *The Journal of Peasant Studies*, vol. 36, no. 1, pp. 139–169, 2009.

6. S.W. Mintz, *Sweetness and Power*. New York: Viking, 1985.

7. M. Nestle, *Food Politics: How the Food Industry Influences Nutrition and Health*. Berkeley, CA: University of California Press, 2013.

8. D.E. Sutton, *Remembrance of Repasts: An Anthropology of Food and Memory*. Oxford: Berg, 2001.

9. D.E. Sutton, "Food and the senses," *Annual Review of Anthropology*, vol. 39, pp. 209–223, 2010.

10. G.N. Dixon and C.E. Clarke, "Heightening uncertainty around certain science media coverage, false balance, and the autism-vaccine controversy," *Science Communication*, vol. 35, no. 3, pp. 358–382, 2013.

11. D.W. Gade, *Nature and Culture in the Andes*. Madison, WI: University of Wisconsin Press, 1999.

12. H. Marvin, *Good to Eat: Riddles of Food and Culture*. Waveland Press, 1998.

13. F.J. Simoons, *Eat Not This Flesh: Food Avoidances from Prehistory to the Present*. Madison, WI: University of Wisconsin Press, 1994.

14. F.J. Simoons, *Plants of Life, Plants of Death*. Madison, WI: University of Wisconsin Press, 1998.

15. R. Aunger, "Are food avoidances maladaptive in the Ituri Forest of Zaire?" *Journal of Anthropological Research*, vol. 50, no. 3, pp. 277–310, 1994.

16. E. Herrmann, J. Call, M.V. Hernández-Lloreda, B. Hare, and M. Tomasello, "Humans have evolved specialized skills of social cognition: The cultural intelligence hypothesis," *Science*, vol. 317, no. 5843, pp. 1360–1366, 2007.

17. A. Appadurai, "How to make a national cuisine: Cookbooks in contemporary India," *Comparative Studies in Society and History*, vol. 30, no. 1, pp. 3–24, 1988.

18. A.B. Bugge and R. Almås, "Domestic dinner: Representations and practices of a proper meal among young suburban mothers," *Journal of Consumer Culture*, vol. 6, no. 2, pp. 203–228, 2006.

19. K. Cairns, J. Johnston, and S. Baumann, "Caring about food: Doing gender in the foodie kitchen," *Gender & Society*, vol. 24, no. 5, pp. 591–615, 2010.

20. M.O. Jones, "Food choice, symbolism, and identity: Bread and butter issues for folkloristics and nutrition studies (American Folklore Society presidential address, October 2005)," *Journal of American Folklore*, vol. 120, no. 476, pp. 129–177, 2007.

21. M. Kahn, *Always Hungry, Never Greedy: Food and the Expression of Gender in a Melanesian Society*. Illinois: Waveland Press, 1993.

22. M.J. Weismantel, *Food, Gender, and Poverty in the Ecuadorian Andes*. Philadelphia: University of Pennsylvania Press, 1989.

23. G. Blue, "Food, publics, science," *Public Understanding of Science*, vol. 19, no. 2, pp. 147–154, 2010.

24. A.M. Collins, "The global farms race: Land grabs, agricultural investment, and the scramble for food security," *Journal of Peasant Studies*, vol. 40, no. 5, pp. 903–906, 2013.

25. J. Jing, ed., *Feeding China's Little Emperors: Food, Children, and Social Change*. Stanford, CA: Stanford University Press, 2000.

26. C. Lentz, ed., *Changing Food Habits: Case Studies from Africa, South America and Europe*. Netherlands: OPA Associates, 1999.

27. H.A. Levenstein, *Paradox of Plenty: A Social History of Eating in Modern America*. Berkeley: University of California Press, 2003.

28. S. Yearley, M. Kugelman, and S.L. Levenstein, The global farms race: Land grabs, agricultural investment, and the scramble for food security," *Food Security*, vol. 5, no. 4, pp. 613–614, 2013.

29. M.A.K. Matossian. *Poisons of the Past: Molds, Epidemics, and History*. Yale University Press, 1989.

30. A.J. Ramos, V. Sanchis, and S. Marin, "The prehistory of mycotoxins: Related cases from ancient times to the discovery of aflatoxins," *World Mycotoxin Journal*, vol. 4, no. 2, pp. 101–112, 2011.

31. D. Sutton, N. Naguib, L. Vournelis, and M. Dickinson, "Food and contemporary protest movements," *Food, Culture and Society: An International Journal of Multidisciplinary Research*, vol. 16, no. 3, pp. 345–366, 2013.

32. N.L. Etkin. *Foods of Association: Biocultural Perspectives on Foods and Beverages that Mediate Sociability*. Tucson, AZ: University of Arizona Press, 2009.

33. M. Nestle and W.A. McIntosh, "Writing the food studies movement," *Food, Culture & Society*, vol. 13, no. 2, pp. 159–168, 2010.

34. J. Allouche, "The sustainability and resilience of global water and food systems: Political analysis of the interplay between security, resource scarcity, political systems and global trade," *Food Policy*, vol. 36, pp. S3–S8, 2011.

35. H.J. Brinkman and C.S. Hendrix. (2011). Food Insecurity and Violent Conflict: Causes: Consequences, and Addressing the Challenges. World Food Programme. Available: http://ucanr.edu/blogs/food2025/blogfiles/14415.pdf. Accessed on June 26, 2015.

36. M.E. Chatwin, *Socio-Cultural Transformation and Foodways in the Republic of Georgia*. Commack, NY: Nova Science, 1997.

37. E. Messer, M.J. Cohen, and J. D'Costa, *Food from Peace: Breaking the Links Between Conflict and Hunger*. Washington: International Food Policy Research Institute, 1998.

38. N. Nunn and N. Qian, (2012). *Aiding Conflict: The Impact of US Food Aid on Civil War*. National Bureau of Economic Research. Available: http://www.nber.org/papers/w17794. Accessed on June 26, 2015.

39. K. Ogden, "Coping strategies developed as a result of social structure and conflict: Kosovo in the 1990s," *Disasters*, vol. 24, no. 2, pp. 117–132, 2000.

40. A. Beardsworth and T. Keil, *Sociology on the Menu: An Invitation to the Study of Food and Society*. London: Routledge, 2013.

41. C.W. Bynum, "Fast, feast, and flesh: The religious significance of food to medieval women," in *Food and Culture: A Reader*, C. Counihan and Penny Van Esterik, Eds. New York: Routledge, 2012, pp.138–158.

42. M. Douglas. *Purity and Danger*. London: Routledge, 2002.

43. N. Fiddes, "Social aspects of meat eating," *Proceedings of the Nutrition Society*, vol. 53, no. 2, pp. 271–279, 1994.

44. K.A. Johnson, A.E. White, B.M. Boyd, and A.B. Cohen. "Matzah, meat, milk, and mana: Psychological influences on religio-cultural food practices," *Journal of Cross-Cultural Psychology*, vol. 42, no. 8, pp. 1421–1436, 2011.

45. C. Lévi-Strauss, "Culinary triangle," *New Society*, vol. 8, no. 221, pp. 937–940, 1966.

46. C. Levi-Strauss, *The Raw and the Cooked*. Octagon Books, 1979.

47. J. Rosenblum, *Food and Identity in Early Rabbinic Judaism*. Cambridge University Press, 2010.

48. M. DeSoucey, "Gastronationalism food traditions and authenticity politics in the European Union," *American Sociological Review*, vol. 75, no. 3, pp. 432–455, 2010.

49. C. Fischler, "Food, self, and identity." *Social Science Information*, vol. 27, no., pp. 275–292, 1988.

50. N. Fox and K.J. Ward, "You are what you eat? Vegetarianism, health and identity," *Social Science and Medicine*, vol. 66, no. 12, pp. 2585–2595, 2008.

51. J.L. Lusk, "The political ideology of food," *Food Policy*, vol. 37, no. 5, pp. 530–542, 2012.

52. M. Montanari, *Italian Identity in the Kitchen, or Food and the Nation*. Columbia University Press, 2013.

53. M.B. Ruby and S.J. Heine, "Meat, morals, and masculinity," *Appetite*, vol. 56, no. 2, pp. 447–450, 2011.

54. E. Sadalla and J. Burroughs, "Profiles in eating—sexy vegetarians and other diet-based social stereotypes," *Psychology Today*, vol. 15, no. 10, pp. 51–57, 1981.

55. T.R. Alley, L.W. Brubaker, and O.M. Fox, "Courtship feeding in humans?" *Human Nature*, vol. 24, no. 4, pp. 430–443, 2013.

56. L. Miller, P. Rozin, and A.P. Fiske, "Food sharing and feeding another person suggest intimacy: Two studies of American college students," *European Journal of Social Psychology*, vol. 28, no. 3, pp. 423–436, 1988.

57. J.W. Traphagan and L.K. Brown, "Fast food and intergenerational commensality in Japan: New styles and old patterns," *Ethnology*, vol. 41, no. 2, pp. 119–134, 2002.

58. W.K. Hallman, "Communicating about microbial risks in foods," in *Microbial Risk Analysis of Foods*, D.W. Schaffner, Ed., Washington, DC: American Society for Microbiology Press, 2008, pp. 205–262.

59. C. Fife Schaw and G. Rowe. "Public perceptions of everyday food hazards: A psychometric study," *Risk analysis*, vol. 16, no. 4, pp. 487–500, 1996.

60. L.J. Frewer, K. Bergmann, M. Brennan, R. Lion, R. Meertens, G. Rowe, M. Siegrist, and C. Vereijken, "Consumer response to novel agri-food technologies: Implications for predicting consumer acceptance of emerging food technologies," *Trends in Food Science & Technology*, vol. 22, no. 8, pp. 442–456, 2011.

61. L. Frewer and A. Fischer, "The evolution of food technology, novel foods, and the psychology of novel food 'acceptance," in *Nanotechnologies in Food*, Q. Chaudry, L. Castle, and R. Watkins, Eds. Cambridge: Royal Society of Chemistry Publishing, 2010, pp. 18–35.

62. L.J. Frewer, C. Howard, and R. Shepherd, "Public concerns in the United Kingdom about general and specific applications of genetic engineering: Risk, benefit, and ethics," *Science, Technology & Human Values*, vol. 22, no. 1, pp. 98–124, 1997.

63. P. Sparks and R. Shepherd, "Public perceptions of the potential hazards associated with food production and food consumption: An empirical study," *Risk Analysis*, vol. 14, no. 5, pp. 799–806, 1994.

64. E. Van Kleef, L.J. Frewer, G.M. Chryssochoidis, J.R. Houghton, S. Korzen-Bohr, T. Krystallis, J. Lassen, U. Pfenning, and G. Rowe, "Perceptions of food risk management among

key stakeholders: Results from a cross-European study," *Appetite*, vol. 47, no. 1, pp. 46–63, 2006.

65. K.K. Jensen, J. Lassen, P. Robinson, and P. Sandøe, "Lay and expert perceptions of zoonotic risks: Understanding conflicting perspectives in the light of moral theory," *International Journal of Food Microbiology*, vol. 99, no. 3, pp. 245–255, 2005

66. P. Slovic, "Perception of risk," *Science*, vol. 236, no. 4799, pp. 280–285, 1987.

67. A.V. Cardello, "Consumer concerns and expectations about novel food processing technologies: Effects on product liking," *Appetite*, vol. 40, no. 3, pp. 217–233, 2003.

68. P. Slovic, "Perceptions of risk: Reflections on the psychometric paradigm," in *Social Theories of Risk*, D. Golding and S. Krimsky, Eds. Westport, CT: Greenwood, 1992, pp. 117–152.

69. C. Cormick, "Lies, deep fries, and statistics!! The search for the truth between public attitudes and public behavior towards genetically modified foods," *Choices*, vol. 20, no. 4, pp. 227–231, 2005.

70. B. Schnettler, G. Crisóstomo, J. Sepúlveda, M. Mora, G. Lobos, H. Miranda, and K.G. Grunert, "Food neophobia, nanotechnology and satisfaction with life," *Appetite*, vol. 69, pp. 71–79, 2013.

71. A.J. Cook and J.R. Fairweather, "Intentions of New Zealanders to purchase lamb or beef made using nanotechnology," *British Food Journal*, vol. 109, no. 9, pp. 675–688, 2007.

72. O.V. Crowley, J. Marquette, D. Reddy, and R. Fleming, "Factors predicting likelihood of eating irradiated meat," *Journal of Applied Social Psychology*, vol. 43, no. 1, pp. 95–105, 2013.

73. J.A. Fox, D.J. Hayes, and J.F. Shogren, "Consumer preferences for food irradiation: How favorable and unfavorable descriptions affect preferences for irradiated pork in experimental auctions," *Journal of Risk and Uncertainty*, vol. 24, no. 1, pp. 75–95, 2002.

74. A. Ronteltap, J.C.M. Van Trijp, R.J. Renes, and LJ. Frewer, "Consumer acceptance of technology-based food innovations: Lessons for the future of nutrigenomics," *Appetite*, vol. 49, no. 1, pp. 1–17, 2007.

75. American Meat Institute Foundation, (2006). Carbon monoxide meat packaging system offers key food safety benefits, new university analyses show. Available: http://www.meatsafety.org/ht/a/GetDocumentAction/i/2165. Accessed on June 26, 2015.

76. T. Boyle, "Groups protest use of carbon monoxide in meat packaging," *USA Today*, 2006 [Online]. Available: http://www.usatoday.com/news/health/2006-02-21-carbon-monoxide -meat_x.htm.

77. Consumer Federation of America, (2006). Most Consumers Are Concerned About Practice of Adding Carbon Monoxide to Meat, New Survey Finds. Available: http://www.consume rfed.org/pdfs/CO_Meat_Consumer_Press_Release_9.25.06.pdf. Accessed on June 26, 2015.

78. R. Weiss, "FDA is urged to ban carbon-monoxide-treated meat," *Washington Post*, February 20, 2006 [Online]. Available: http://www.washingtonpost.com/wp-dyn/content/article/ 2006/02/19/AR2006021901101.html

79. C. Grebitus, H. Jensen, J. Roosen, and J. Sebranek, "Consumer Acceptance of Fresh Meat Packaging with Carbon Monoxide," Iowa State University Animal Industry Report: AS 659, ASL R2756, 2013. Available: http://lib.dr.iastate.edu/ans_air/vol659/iss1/7. Accessed on June 26, 2015.

80. C. Guilloryt, "Is carbon monoxide in your food?," *Examiner.com*, July 31, 2012. Available: http://www.examiner.com/article/is-carbon-monoxide-your-food.

81. D. Grobe and R. Douthitt, "Consumer acceptance of recombinant bovine growth hormone: Interplay between beliefs and perceived risks," *Journal of Consumer Affairs*, vol. 29, no. 1, pp. 128–143, 1995.

82. W.P. Preston, A.M. McGuirk, and G.M. Jones, "Consumer reaction to the introduction of bovine somatotropin," In *Economics of Food Safety*, J. Caswell, Ed., Netherlands: Springer Netherlands, 1991, pp. 189–210.

83. American Cancer Society., (2014). Recombinant bovine growth hormone. Available: http://www.cancer.org/cancer/cancercauses/othercarcinogens/athome/recombinant-bovine -growth-hormone. Accessed on June 26, 2015.

84. Food and Drug Administration (FDA), (2014). Bovine somatotropin. Available: http://www .fda.gov/AnimalVeterinary/SafetyHealth/ProductSafetyInformation/ucm055435.htm. Accessed on June 26, 2015.

85. L. Busch, "Biotechnology: Consumer concerns about risks and values," *Food Technology (USA)*, vol. 45, pp. 96–101, 1991.

86. L. Zepeda, R. Douthitt, and S.Y. You, "Consumer risk perceptions toward agricultural biotechnology, self-protection, and food demand: The case of milk in the United States," *Risk Analysis*, vol. 23, no. 5, pp. 973–984, 2003.

87. S.G. Hadden, *A Citizen's Right to Know: Risk Communication and Public Policy*. Boulder, CO: Westview Press, 1989.

88. S.K. Harlander, "Social, moral, and ethical issues in food biotechnology," *Food Technology*, vol. 45, no. 5, pp. 152–161, 1991.

89. Food and Water Watch (2014). Bovine growth hormone. Available: https://www.foodandwa terwatch.org/food/foodsafety/dairy/. Accessed on June 26, 2015.

90. G. Gaskell, N. Allum, M. Bauer, J. Jackson, S. Howard, and N.M. Lindsey, "Ambivalent GM nation? Public attitudes to biotechnology in the UK, 1991–2002," Life Sciences in European Society Report: London School of Economics and Political Science, 2003. Available: file:///C:/Users/mnucci/Downloads/d912f509a198a7ae21%20(1).pdf.

91. M.W. Bauer, "Distinguishing red and green biotechnology: Cultivation effects of the elite press," *International Journal of Public Opinion Research*, vol. 17, no. 1 pp. 63–89, 2005.

92. W.K. Hallman, A. Adelaja, B. Schilling, and J. Lang, "Public perceptions of genetically modified food: Americans know not what they eat," Food Policy Institute Report RR-0302—001, 2002. Available: http://johnlang.org/pubs/NationalStudy2002.pdf. Accessed on June 26, 2015.

93. W.K. Hallman, W. C. Hebden, H. Aquino, C. Cuite, and J. Lang, "Public perceptions of genetically modified foods: A national study of American knowledge and opinion," Food Policy Institute Report RR-1003–004, 2003. Available: http://foodpolicy.rutgers.edu/docs/pubs/2003_Public_Perceptions_of_Genetically_Modified_Foods.pdf. Accessed on June 26, 2015.

94. W.K. Hallman, C. Hebden, C. Cuite, H. Aquino, and J. Lang, "Americans and GM food: Knowledge, opinion and interest in 2004," Food Policy Institute report RR-1104 –007, 2004. Available: http://foodpolicy.rutgers.edu/docs/pubs/2004_Americans%20and %20GM%20Food_Knowledge%20Opinion%20&%20Interest%20in%202004.pdf. Accessed on June 26, 2015.

95. W. Hallman and J. Metcalfe, (1993). Public perceptions of agricultural biotechnology: A survey of New Jersey residents. New Jersey Agricultural Experiment Station, Rutgers,

the State University of New Jersey, New Brunswick NJ. Available: http://ageconsearch .umn.edu/bitstream/18170/1/pa94ha01.pdf. Accessed on June 26, 2015.

96. B. Onyango, R. Govindasamy and R. Nayga, Food Policy Institute, Working Paper No. WP1104–017. (2004). Measuring US consumer preferences for Genetically Modified Foods using Choice Modeling Experiments: The Role of Price, Product Benefits and Technology. Available: http://ideas.repec.org/p/ags/rutfwp/18181.html. Accessed on June 26, 2015.

97. J.L. Brown and Y. Ping, "Comparison of consumer reaction to information about two genetically engineered soybeans that differ in consumer benefit," *Journal of International Food & Agribusiness Marketing*, vol. 13, no. 1, pp. 7–25, 2003.

98. L.J. Frewer, I.A. van der Lans, A.R.H. Fischer, M.J. Reinders, D. Menozzi, X. Zhang, I. van den Berg, and K.L. Zimmermann, "Public perceptions of agri-food applications of genetic modification–a systematic review and meta-analysis," *Trends in Food Science and Technology*, vol. 30, no. 2, pp. 142–152, 2013.

99. N. Gupta, A.R.H. Fischer, and L.J. Frewer, "Socio-psychological determinants of public acceptance of technologies: A review," *Public Understanding of Science*, vol. 21, no. 7, pp. 782–795, 2012.

100. S.S. Ho, D.A. Scheufele, and E.A. Corley, "Factors influencing public risk–benefit considerations of nanotechnology: Assessing the effects of mass media, interpersonal communication, and elaborative processing," *Public Understanding of Science*, vol. 22, no. 5, pp. 606–623, 2013.

101. F. Rollin, J. Kennedy, and J. Wills, "Consumers and new food technologies," *Trends in Food Science & Technology*, vol. 22, no. 2, pp. 99–111, 2011.

102. M. Siegrist, "The influence of trust and perceptions of risks and benefits on the acceptance of gene technology," *Risk Analysis*, vol. 20, no. 2, pp. 195–204, 2000.

103. L. Bredahl, "Determinants of consumer attitudes and purchase intentions with regard to genetically modified food–results of a cross-national survey," *Journal of Consumer Policy*, vol. 24, no. 1, pp. 23–61, 2001.

104. J.F.M. Swinnen and T. Vandemoortele, (2011). On butterflies and Frankenstein: A dynamic theory of regulation. LICOS Centre for Institutions and Economic Performance. Available: http://www.econstor.eu/bitstream/10419/74958/1/dp276.pdf.

105. G. Gaskell, M.W. Bauer, J. Durant, and N.C. Allum, "Worlds apart? The reception of genetically modified foods in Europe and the US," *Science*, vol. 285, no. 5426, pp. 384–387, 1999.

106. S. Ceccoli and W. Hixon, "Explaining attitudes toward genetically modified foods in the European Union," *International Political Science Review*, vol. 33, no. 3, pp. 301–319, 2012.

107. F. Hossain, B. Onyango, A. Adelaja, B. Schilling, and W. Hallman, "Consumer acceptance of food biotechnology: Willingness to buy genetically modified food products," *Journal of International Food and Agribusiness Marketing*, vol. 15, no. 1–2, pp. 53–76, 2004.

108. P. Macnaghten, M.B. Kearnes, and B. Wynne, "Nanotechnology, governance, and public deliberation: What role for the social sciences?" *Science Communication*, vol. 27, no. 2, pp. 268–291, 2005.

109. W.K. Hallman and S. Condry, (2006). Public Opinion and Media Coverage of Animal Cloning and the Food Supply: Executive summary. Food Policy Institute. Publication No. RR-1106–011. Available: http://foodpolicy.rutgers.edu/docs/pubs/Animal%20Cloning %202006.pdf. Accessed on June 26, 2015.

110. D. Nelkin, "An uneasy relationship: The tensions between medicine and the media," *The Lancet*, vol. 347, no. 9015, pp. 1600–1603, 1996.

111. National Science Board (2012). Science and Engineering Indicators 2012. Arlington, VA: (NSB 10–01). National Science Foundation. Available: http://www.nsf.gov/statistics/seind12/.

112. D.A. Scheufele, "Messages and heuristics: How audiences form attitudes about emerging technologies," in *Engaging Science: Thoughts, Deeds, Analysis and Action*, J. Turney, Ed. London: The Wellcome Trust, 2006, pp. 20–25.

113. M.C. Nisbet, D. Brossard, and A. Kroepsch, "Framing science the stem cell controversy in an age of press/politics," *The International Journal of Press/Politics*, vol. 8, no. 2, pp. 36–70, 2003.

114. S.H. Stocking and L.W. Holstein, "Manufacturing doubt: Journalists' roles and the construction of ignorance in a scientific controversy," *Public Understanding of Science*, vol. 18, no. 1, pp. 23–42, 2009.

115. C. Chang, "Men's and women's responses to two-sided health news coverage: A moderated mediation model," *Journal of Health Communication*, vol. 18, no. 11, pp. 1326–1344, 2013.

116. R. Bayer, D. Merritt Johns, and S. Galea, "Salt and public health: Contested science and the challenge of evidence-based decision making," *Health Affairs*, vol. 31, no. 12, pp. 2738–2746, 2012.

117. C.P. Cooper and D. Yukimura, "Science writers' reactions to a medical "breakthrough story," *Social Science and Medicine*, vol. 54, no. 12, pp. 1887–1896, 2002.

118. S. Dentzer, "Communicating medical news—pitfalls of health care journalism," *New England Journal of Medicine*, vol. 360, no. 1, pp. 1–3, 2009.

119. K.J. Eskine, "Wholesome foods and wholesome morals? Organic foods reduce prosocial behavior and harshen moral judgments," *Social Psychological and Personality Science*, vol. 4, no. 2, pp. 251–254, 2013.

120. J.D. Jensen, "Scientific uncertainty in news coverage of cancer research: Effects of hedging on scientists' and journalists' credibility," *Human Communication Research*, vol. 34, no. 3, pp. 347–369, 2008.

121. M.G. Pellechia, "Trends in science coverage: A content analysis of three US newspapers," *Public Understanding of Science*, vol. 6, no. 1, pp. 49–68, 1997.

122. S.H. Stocking, "How journalists deal with scientific uncertainty," in *Communicating Uncertainty: Media Coverage of New and Controversial Science*. S. Friedman, S. Dunwoody, and C. Rogers, Eds. Mahwah, NJ: Lawrence Erlbaum, 1999, pp. 23–42.

123. R. Coleman, E. Thorson, and L. Wilkins, "Testing the effect of framing and sourcing in health news stories," *Journal of Health Communication*, vol. 16, no. 9, pp. 941–954, 2011.

124. L.A. Marks, N. Kalaitzandonakes, L. Wilkins, and L. Zakharova, "Mass media framing of biotechnology news," *Public Understanding of Science*, vol. 16, no. 2, pp. 183–203, 2007.

125. D. Tewksbury and D. Scheufele, "News frames theory and research," in *Media effects. Advances in Theory and Research*. J. Bryant and M.B. Oliver, Eds. New York: Lawrence Erlbaum Associates, Inc., 2009, pp. 17–33.

126. Z. Pan and G.M. Kosicki, "Framing analysis: An approach to news discourse," *Political Communication*, vol. 10, no. 1, pp. 55–75, 1993.

127. S. Hornig, "Science stories: Risk, power and perceived emphasis, "*Journalism & Mass Communication Quarterly*, vol. 67, no. 4, pp. 767–776, 1990.

128. B.M. Miller, A.A. Packer, and B. Barnett, "Reporting risk: Perceptions of fear and risk from health news coverage," *Communication Research Reports*, vol. 28, no. 3, pp. 244–253, 2011.

129. T.A. Ten Eyck and M. Williment, "The national media and things genetic coverage in the *New York Times* (1971–2001) and the *Washington Post* (1977-2001)," *Science Communication*, vol. 25, no. 2, pp. 129–152, 2003.

130. G. Tuchman, *Making News: A Study in the Construction of Reality*. New York: Free Press, 1978.

131. M. Hivon, P. Lehoux, J-L. Denis, and M. Rock, "Marginal voices in the media coverage of controversial health interventions: How do they contribute to the public understanding of science?" *Public Understanding of Science*, vol. 19, no. 1, pp. 34–51, 2010.

132. T. Listerman, "Framing of science issues in opinion-leading news: International comparison of biotechnology issue coverage," *Public Understanding of Science*, vol. 19, no. 1, pp. 5–15, 2010.

133. M.C. Nisbet and B. V. Lewenstein, "Biotechnology and the American media the policy process and the Elite Press, 1970 to 1999," *Science Communication*, vol. 23, no. 4, pp. 359–391, 2002.

134. M.C. Nisbet and M. Huge, "Attention cycles and frames in the plant biotechnology debate managing power and participation through the press/policy connection," *The Harvard International Journal of Press/Politics*, vol. 11, no. 2 pp. 3–40, 2006.

135. T.A. Ten Eyck, P.B. Thompson, and S.H. Priest (2001). "Biotechnology in the United States of America: Mad or moral science?" in *Biotechnology 1996–2000: The years of controversy*, G. Gaskell and M.W. Bauer, Eds. London: NMSI Trading Ltd., Science Museum, 2001, pp. 307–318.

136. L.P. Plein, "Popularizing biotechnology: The influence of issue definition," *Science, Technology & Human Values*, vol. 16, no. 4, pp. 474–490, 1991.

137. S.H. Priest, *A Grain of Truth: The Media, the Public, and Biotechnology*. Maryland: Rowman & Littlefield Publishers, 2002.

138. S.H. Priest and J. Talbert, "Mass media and the ultimate technological fix: Newspaper coverage of biotechnology," *Southwestern Mass Communication Journal*, vol. 10, no. 1, pp. 76–85, 1994.

139. S.H. Priest and A.W. Gillespie, "Seeds of discontent: Expert opinion, mass media messages, and the public image of agricultural biotechnology," *Science and Engineering Ethics*, vol. 6, no. 2, pp. 529–39, 2000.

140. B.H. Reber and B.K. Berger, "Framing analysis of activist rhetoric: How the Sierra Club succeeds or fails at creating salient messages," *Public Relations Review*, vol. 31, no. 2, pp. 185–195, 2005.

141. A.G. Anderson, "The media politics of oil spills," *Spill Science & Technology Bulletin*, vol. 7, no. 1, pp. 7–15, 2002.

142. A. Anderson, "Source-media relations: The production of the environmental agenda," in *The Mass Media and Environmental Issues*, A. Hansen, Ed. New York, NY: Leicester University Press, 1993, pp. 51–68.

143. K. Carmody, "It's a jungle out there," *Columbia Journalism Review*, vol. 34, no. 1, pp. 40–44, 1995.

144. J. Cracknell, "Issue arenas, pressure groups and environmental agendas," in *The Mass Media and Environmental Issues*, A. Hansen, Ed. New York: Leicester University Press, 1993, pp. 51–68.

145. S. Kwanl "Framing the fat body: Contested meanings between government, activists, and industry," *Sociological Inquiry*, vol. 79, no. 1, pp. 25–50, 2009.

146. J.D. McCarthy, J. Smith, and M.N. Zald, "Assessing public media, electoral, and governmental agendas," in *Comparative Perspectives on Social Movements Opportunities, Mobilizing Structures, and Framing*, D. McAdam, J.D. McCarthy, and M.N. Zald, Eds. Cambridge, UK: Cambridge University Press, 1996, pp. 291–311.

147. T. Baylor, "Media framing of movement protest: The case of American Indian protest," *The Social Science Journal*, vol. 33, no. 3, pp. 241–255, 1996.

148. B. Klandermans and S. Goslinga, "Media discourse, movement publicity, and the generation of collective action frames: Theoretical and empirical exercises in meaning construction," in *Comparative Perspectives on Social Movements: Political Opportunities, Mobilizing Structures, and Cultural Framings*, D. McAdam, J.D. McCarthy, and M.N. Zald, Eds. Cambridge: Cambridge University Press, 1996, pp. 312–337.

149. L.A. Marks and N. Kalaitzandonakes, "Mass media communications about agrobiotechnology," *AgBioForum*, vol. 4, no. 3&4, pp. 199–208, 2001.

150. M.L. Nucci and R. Kubey, ""We Begin Tonight with Fruits and Vegetables: Genetically modified food on the evening news 1980–2003," *Science Communication*, vol. 29, no. 2, pp. 147–176, 2007.

151. S.H. Priest, "Information equity, public understanding of science, and the biotechnology debate," *Journal of Communication*, vol. 45, no. 1, pp. 39–54, 1995.

152. L. Marks, "Communicating about agrobiotechnology," *AgBioForum*, vol. 4, no. 3, pp. 152–154, 2001.

153. L.A. Marks, N.G. Kalaitzandonakes, and S.S. Vickner, "Evaluating consumer response to GM foods: Some methodological considerations," *CAFRI: Current Agriculture, Food and Resource Issues*, vol. 04, pp. 80–94, 2003.

154. C. McInerney, N. Bird, and M. Nucci, "The flow of scientific knowledge from lab to the lay public: The case of genetically modified food," *Science Communication*, vol. 26, no. 1, pp. 44–74, 2004.

155. K. Hart, *Eating in the Dark: America's Experiment with Genetically Modified Food*. New York: Pantheon, 2007.

156. Pew Initiative on Food and Biotechnology, (2002). Three Years Later: Genetically Engineered Corn and the Monarch Butterfly Controversy. Available: http://www.pewtrusts.org/en/research-and-analysis/reports/2002/06/10/three-years-later-genetically-engineered-corn-and-the-monarch-butterfly-controversy. Accessed on June 26, 2015.

157. M.L. Winston, *Travels in the Genetically Modified Zone*. Cambridge, MA: Harvard University Press, 2002.

158. J.E. Losey, L.S. Rayor, and M.E. Carter, "Transgenic pollen harms monarch larvae," *Nature*, vol. 399, no. 6733, pp. 214–214, 1999.

159. M. Nisbet, Matt, and B.V. Lewenstein, "A comparison of US media coverage of biotechnology with public perceptions of genetic engineering 1995–1999," *2001 International Public Communication of Science and Technology Conference*, Geneva. 2001.

160. B. Plumer "Monarch butterflies keep disappearing. Here's why," *Washington Post*, January 29, 2014 [Online]. Available: http://www.washingtonpost.com/blogs/wonkblog/wp/2014/01/29/the-monarch-butterfly-population-just-hit-a-record-low-heres-why/.

161. W.A. Gamson and A. Modigliani, "Media discourse and public opinion on nuclear power: A constructionist approach," *American Journal of Sociology*, vol. 95, no. 1, pp. 1–37, 1989.

162. ETC Group (2004). Down on the farm: The impact of nano-scale technologies on food and agriculture. Available: http://www.etcgroup.org/content/down-farm-impact-nano-scale-technologies-food-and-agriculture.

163. CSREES (2010). Nanotechnology. Available: http://www.nifa.usda.gov/ProgViewOverview.cfm?prnum=16500.

164. T.V. Duncan, "Applications of nanotechnology in food packaging and food safety: Barrier materials, antimicrobials and sensors," *Journal of Colloid and Interface Science*, vol. 363, no. 1, pp. 1–24, 2011.

165. A. Prakash, S. Sen, and R. Dixit, "The emerging usage and applications of nanotechnology in food processing industries: The new age of nanofood," *International Journal of Pharmaceutical Sciences Review & Research*, vol. 22, no. 1, pp. 107–111, 2013.

166. J.W. Rhim, H.M. Park, and C.S. Ha, "Bio-nanocomposites for food packaging applications," *Progress in Polymer Science*, vol. 38, no. 10, pp. 1629–1652, 2013.

167. M. Avella, R. Avolio, E. Di Pace, M.E. Errico, G. Gentile, and M.G. Volpe, "Polymer-based nanocomposites for food packaging applications," in *Bio-Nanotechnology: A Revolution in Food, Biomedical and Health Sciences*, D. Bagchi, M. Bagchi, H. Moriyama, F. Shahidi, Eds. Wiley-Blackwell, 2013, pp. 212–226.

168. J. Weiss, P. Takhistov, and D. J. McClements, "Functional materials in food nanotechnology," *Journal of Food Science*, vol. 71, no. 9, pp. R107–R116, 2006.

169. V.L. Hanson, "Envisioning ethical nanotechnology: The rhetorical role of visions in postponing societal and ethical implications research," *Science as Culture*, vol. 20, no. 1, pp. 1–36, 2011.

170. R. Shoemaker, D.D. Johnson, and E. Golan, "Consumers and the future of biotech foods in the United States," *Amber Waves*, vol. 1, no. 5, pp. 30–36, 2003.

171. A. Dudo, D.H. Choi, and D.A. Scheufele, "Food nanotechnology in the news: Coverage patterns and thematic emphases during the last decade," *Appetite*, vol. 56, no. 1, pp. 78–89, 2011.

172. M. Gibbons, "Science's new social contract with society," *Nature*, vol. 402, no. 6761, pp. C81–C84, 1999.

173. D. Nelkin. *Selling Science: How the Press Covers Science and Technology*. New York: Freeman, 1995.

174. M.W. Bauer, J. Durant, and G. Gaskell, *Biotechnology in the Public Sphere: A European Sourcebook*. London: Science Museum Press, 1998.

175. S.M. Friedman and B.P. Egolf, "Nanotechnology: Risks and the media," *IEEE Technology and Society Magazine*, vol. 24, no. 4, pp. 5–11, 2005.

176. S.M. Friedman and B. P. Egolf (2007). Woodrow Wilson Center for International Scholars. Changing patterns of mass media coverage of nanotechnology risks. Available: http://mail.nanotechproject.org/process/assets/files/5920/woodrowwilsonwashingtondec182007final.pdf.

177. D.A. Weaver and B. Bimber, "Finding news stories: A comparison of searches using Lexis-Nexis and Google News," *Journalism and Mass Communication Quarterly*, vol. 85, no. 3, pp. 515–530, 2008.

178. G. Gaskell, T. Ten Eyck, J. Jackson, and G. Veltri, "From our readers: Public attitudes to nanotech in Europe and the United States," *Nature Materials*, vol. 3, no. 8, pp. 496–496, 2004.

179. B.V. Lewenstein, J. Radin, and J. Diels, "Nanotechnology in the media: A preliminary analysis," in *Nanotechnology: Societal Implications II: Individual Perspective*, M.C. Roco and W.S. Bainbridge. Dordrecht: Springer, 2004, pp. 258–265.

180. L.F. Stephens, "News narratives about nano S&T in major US and non-US newspapers," *Science Communication*, vol. 27, no. 2, pp. 175–199, 2005.

181. A. Anderson, S. Allan, A. Petersen, and C. Wilkinson, "The framing of nanotechnologies in the British newspaper press," *Science Communication*, vol. 27, no. 2, pp. 200–220, 2005.

182. C. Wilkinson, S. Allan, A. Anderson, and A. Petersen, "From uncertainty to risk? Scientific and news media portrayals of nanoparticle safety," *Health, Risk & Society*, vol. 9, no. 2, pp. 145–157, 2007.

183. B.V. Lewenstein, J. Gorss, and J. Radin (2005). The salience of small: Nanotechnology coverage in the American press, 1986–2004. Available: https://ecommons.library.cornell.edu/bitstream/1813/14275/2/LewensteinGorssRadin.2005.NanoMedia.ICA.pdf.

184. P.J.A. Borm, "Particle toxicology: From coal mining to nanotechnology," *Inhalation Toxicology*, vol. 14, no. 3, pp. 311–324, 2002.

185. R.D. Handy and B.J. Shaw, "Toxic effects of nanoparticles and nanomaterials: Implications for public health, risk assessment and the public perception of nanotechnology," *Health, Risk and Society*, vol. 9, no. 2, pp. 125–144, 2007.

186. G. Oberdörster, E. Oberdörster, and J. Oberdörster, "Nanotoxicology: An emerging discipline evolving from studies of ultrafine particles," *Environmental Health Perspectives*, vol. 113, no. 7, pp. 823–839.

187. D.A. Weaver, E. Lively, and B. Bimber, "Searching for a frame news media tell the story of technological progress, risk, and regulation," *Science Communication*, vol. 31, no. 2, pp. 139–16, 2009.

188. W.S. Bainbridge, "Public attitudes toward nanotechnology," *Journal of Nanoparticle Research*, vol. 4, no. 6, pp. 561–570, 2002.

189. J. Macoubrie, "Public perceptions about nanotechnology: Risks, benefits and trust," *Journal of Nanoparticle Research*, vol. 6, no. 4, pp. 395–405, 2004.

190. G. Gaskell, T. Ten Eyck, J. Jackson, and G. Veltri, "Imagining nanotechnology: Cultural support for technological innovation in Europe and the United States," *Public Understanding of Science*, vol. 14, no. 1, pp. 81–90, 2005.

191. C.J. Lee, D.A. Scheufele, and B.V. Lewenstein, "Public attitudes toward emerging technologies examining the interactive effects of cognitions and affect on public attitudes toward nanotechnology," *Science Communication*, vol. 27, no. 2, pp. 240–267, 2005.

192. A. Matin, E. Goddard, F. Vandermoere, S. Blanchemanche, A. Bieberstein, S. Marette, and J. Roosen, "Do environmental attitudes and food technology neophobia affect perceptions of the benefits of nanotechnology?" *International Journal of Consumer Studies*, vol. 36, no. 2, pp. 149–157, 2012.

193. J. Macoubrie (2005). Informed public perceptions of nanotechnology and trust in government. Project on Emerging Nanotechnologies of the Woodrow Wilson International Center for Scholars. Available: http://www.wilsoncenter.org/publication/pen-1-informed-public-perceptions-nanotechnology-and-trust-government.

194. Pew Project on Emerging Technologies, (2005). Informed Public Perceptions of Nanotechnology and Trust in Government. Available: http://www.pewtrusts.org/our_work_report_detail.aspx?id=19674.

195. Pew Project on Emerging Technologies, (2007). Awareness of and attitudes toward nanotechnology and federal regulatory agencies. Available: http://www.pewtrusts.org/our_work _report_detail.aspx?id=30539.

196. D.A. Scheufele, E. A. Corley, T.J. Shih, K.E. Dalrymple, and S.S. Ho, "Religious beliefs and public attitudes toward nanotechnology in Europe and the United States," *Nature nanotechnology*, vol. 4, no. 2, pp. 91–94, 2009.

197. S. Priest, T. Greenhalgh, and V. Kramer, "Risk perceptions starting to shift? US citizens are forming opinions about nanotechnologym," *Journal of Nanoparticle Research*, vol. 12, no. 1, pp. 11–20, 2010.

198. Harris Poll, (2012). Nanotechnology awareness may be low, but opinions are strong. Available: http://www.harrisinteractive.com/NewsRoom/HarrisPolls/tabid/447/ctl/ReadCustom %20Default/mid/1508/ArticleId/1073/Default.aspx.

199. F. Vandermoere, S. Blanchemanche, A. Bieberstein, S. Marette, and J. Roosen, "The public understanding of nanotechnology in the food domain: The hidden role of views on science, technology, and nature," *Public Understanding of Science*, vol. 20, no. 2, pp. 195–206, 2011.

200. S. Priest, T. Lane, T. Greenhalgh, L.J. Hand, and V. Kramer, "Envisioning emerging nanotechnologies: A three-year panel study of South Carolina citizens," *Risk Analysis*, vol. 31, no. 11, pp. 1718–1733, 2011.

201. A. Retzbach, J. Marschall, M. Rahnke, L. Otto, and M. Maier, "Public understanding of science and the perception of nanotechnology: The roles of interest in science, methodological knowledge, epistemological beliefs, and beliefs about science," *Journal of Nanoparticle Research*, vol. 13, no. 12, pp. 6231–6244, 2011.

202. S.C. Currall, E.B. King, N. Lane, J, Madera, and S. Turrner, "What drives public acceptance of nanotechnology?" *Nature Nanotechnology*, vol. 1, no. 3 pp. 153–155, 2006.

203. M. Siegrist, N. Stampfli, H. Kastenholz, and C. Keller, "Perceived risks and perceived benefits of different nanotechnology foods and nanotechnology food packaging," *Appetite*, vol. 51, no. 2, pp. 283–290, 2008.

204. E. López-Vázquez, T.A. Brunner, and M. Siegrist, "Perceived risks and benefits of nanotechnology applied to the food and packaging sector in México," *British Food Journal*, vol. 114, no. 2, pp. 197–205, 2012.

205. M. Fishbein and I. Ajzen, *Belief, Attitude, Intention and Behavior: An Introduction to Theory and Research*. Reading, MA: Addison-Wesley, 1975.

206. E. Townsend and S. Campbell, "Psychological determinants of willingness to taste and purchase genetically modified food," *Risk Analysis*, vol. 24, no. 5, pp. 1385–1393, 2004.

207. D.N. Cox, A. Koster, and C.G. Russell, "Predicting intentions to consume functional foods and supplements to offset memory loss using an adaptation of protection motivation theory," *Appetite*, vol. 43, no. 1, pp. 55–64, 2004.

208. D. Brossard, D.A. Scheufele, E. Kim, and B.V. Lewenstein, "Religiosity as a perceptual filter: Examining processes of opinion formation about nanotechnology," *Public Understanding of Science*, vol. 18, no. 5, pp. 546–558, 2009.

209. M. Siegrist, "Factors influencing public acceptance of innovative food technologies and products," *Trends in Food Science & Technology*, vol. 19, no. 11, pp. 603–608, 2008.

210. D. Brossard and M.C. Nisbet, "Deference to scientific authority among a low information public: Understanding US opinion on agricultural biotechnology," *International Journal of Public Opinion Research*, vol. 19, no. 1, pp. 24–52, 2007.

211. C.J. Lee, and D.A. Scheufele, "The influence of knowledge and deference toward scientific authority: A media effects model for public attitudes toward nanotechnology," *Journalism and Mass Communication Quarterly*, vol. 83, no. 4, pp. 819–834, 2006.

212. M. Siegrist, M.E. Cousin, H. Kastenholz, and A. Wiek, "Public acceptance of nanotechnology foods and food packaging: The influence of affect and trust," *Appetite*, vol. 49, no. 2, pp. 459–466, 2007.

213. M. Siegrist, A. Wiek, A. Helland, and H. Kastenholz, "Risks and nanotechnology: The public is more concerned than experts and industry," *Nature Nanotechnology*, vol. 2, no. 2, pp. 67–67, 2007.

214. A.V.A. Resurreccion, F.C.F. Galvez, S.M. Fletcher, and S.K. Misra, "Consumer attitudes toward irradiated food: Results of a new study," *Journal of Food Protection*, vol. 58, no. 2, pp. 193–196, 1995.

215. Functional Ingredients (2006, September). US sterol foods market growing - but slowly. Available: http://newhope360.com/trends/us-sterol-foods-market-growing-slowly.

216. G.A. Kundahl., "Communications in the Age of Nanotechnology," In *The Yearbook of Nanotechnology in Society, Volume I: Presenting Futures*, E. Fisher, C. Selin, and J.M. Wetmore, Eds. Netherlands: Springer Science+ Business Media BV, 2008, pp. 183–194.

217. National Nanotechnology Initiative, (undated). Nanotechnology: Big things from a tiny world. Available at http://www.nano.gov/node/240.

218. Center for Responsible Nanotechnology, (2008). What is nanotechnology? Available: http://www.crnano.org/whatis.htm.

219. Food and Drug Administration, (2012). Science and Research: Nanotechnology. [Online]. Available: http://www.fda.gov/scienceresearch/specialtopics/nanotechnology/default.htm.

220. K. Kjølberg and F. Wickson, "Social and ethical interactions with nano: Mapping the early literature," *Nanoethics*, vol. 1, no. 2, pp. 89–104, 2007.

221. J. Balbus, R. Denison, K. Florini, and S. Walsh, "Getting nanotechnology right the first time," in *Nanotechnology: Risk. Ethics and Law*. G. Hunt and M. Mehta, Eds. London: Earthscan, 2006, pp. 130–138.

222. J. Barnett, A. Carr and R. Clift, "Going public: Risk, trust and public understanding of nanotechnologies," in *Nanotechnology: Risk. Ethics and Law*, G. Hunt and M. Mehta, Eds. London: Earthscan, 2006, pp. 196–211.

223. K. David and P.B. Thompson, *What Can Nanotechnology Learn from Biotechnology? Social and Ethical Lessons for Nanoscience from the Debate Over Agrifood Biotechnology and GMOs*. Burlington, MA: Academic Press, 2011.

224. W.K. Hallman, "GM foods in hindsight," in *Emerging Technologies: From Hindsight to Foresight*, E. Einsiedel, Ed. Canada: University of British Columbia Press, 2008, pp. 13–32.

225. J. Macoubrie, "Nanotechnology: Public concerns, reasoning and trust in government," *Public Understanding of Science*, vol. 15, no. 2, pp. 221–241, 2006.

226. M.D. Mehta, "From biotechnology to nanotechnology: What can we learn from earlier technologies?" *Bulletin of Science, Technology and Society*, vol. 24, no. 1, pp. 34–39, 2004.

227. J. Schummer, (2006). "Societal and ethical implications of nanotechnology: Meanings, interest groups, and social dynamics," in J. Schummer and D. Baird, Eds. *Nanotechnology Challenges: Implications for Philosophy, Ethics and Society*. Singapore: World Scientific Publishing, 2006, pp. 413–449.

228. J.R. Wolfson, "Social and ethical issues in nanotechnology: Lessons from biotechnology and other high technologies," *Biotechnology Law Report*, vol. 22, no. 4, pp. 376–396, 2003.

229. M.V. Alario and W.W. Freudenburg, "High-risk technology, legitimacy and science: The US search for energy policy consensus," *Journal of Risk Research*, vol. 9, no. 7, pp. 737–753, 2006.

230. I. Hargreaves and G. Ferguson, "Who's misunderstanding whom? An inquiry into the relationship between science and the media," Economic and Social Research Council, 2000.

231. A.L. Peláez and J.A. Díaz. "Science, technology and democracy: Perspectives about the complex relation between the scientific community, the scientific journalist and public opinion," *Social Epistemology*, vol. 21, no. 1, pp. 55–68, 2007.

232. A. Touraine, "The crisis of progress," in *Resistance to New Technology: Nuclear Power, Information Technology, Biotechnology*, M.W. Bauer, Ed. Cambridge: Cambridge University Press, 1985, pp. 45–56. *States of knowledge: the co-production of science and the social order*. Routledge, 2013.

233. P. Sturgis and N. Allum, "Science in society: Re-evaluating the deficit model of public attitudes," *Public Understanding of Science*, vol. 13, no. 1, pp. 55–74, 2004.

234. R. Mulgan, "Perspectives on 'the public interest' Paper prepared for a proposed IPAA (ACT Division) seminar in Sept, 1999.," *Canberra Bulletin of Public Administration*, vol. 95, pp. 5, 2000.

235. S. Jasanoff, *States of Knowledge: The Co-Production of Science and Social Order*. London: Routledge, 2004.

236. Field of Dreams. (1989). Motion Picture: Universal Studios.

237. M.C. Nisbet, "The competition for worldviews: Values, information, and public support for stem cell research," *International Journal of Public Opinion Research*, vol. 17, no. 1, pp. 90–112, 2005.

238. M.C. Nisbet and R.K. Goidel, "Understanding citizen perceptions of science controversy: Bridging the ethnographic-survey research divide," *Public Understanding of Science*, vol. 16, no. 4, pp. 421–440, 2007.

239. S.G. Sapp, "A comparison of alternative theoretical explanations of consumer food safety assessments," *International Journal of Consumer Studies*, vol. 27, no. 1, pp. 34–39, 2003.

240. A.M. Weinberg, "Science and trans-science," *Minerva*, vol. 10, no. 2, pp. 209–222, 1972.

241. N. Allum, P. Sturgis, D. Tabourazi, and I. Brunton-Smith, "Science knowledge and attitudes across cultures: A meta-analysis," *Public Understanding of Science*, vol. 17, no. 1, pp. 35–54, 2008.

242. J.C. Besley and J. Shanahan, "Media attention and exposure in relation to support for agricultural biotechnology," *Science Communication*, vol. 26, no. 4, pp. 347–367, 2005.

243. T. Changthavorn, "Bioethics of IPRs: What does a Thai Buddhist think." Roundtable discussion on Bioethical Issues of IPRs, Selwyn College, University of Cambridge, 2003

244. C. Kachonpadungkitti and D.R.J. Macer, "Attitudes to bioethics and biotechnology in Thailand (1993–2000), and impacts on employment," *Eubios Journal of Asian and International Bioethics*, vol. 14, pp. 118–134, 2004.

245. A.A. Leiserowitz, "American risk perceptions: Is climate change dangerous?" *Risk Analysis*, vol. 25, no. 6, pp. 1433–1442, 2005.

246. P. Slovic and E. Peters, "The importance of worldviews in risk perception," *Risk Decision and Policy*, vol. 3, no. 2, pp. 165–170, 1998.

247. D.M. Kahan, D. Braman, and G. N. Mandel (2009). Risk and culture: Is Synthetic Biology Different? Harvard Law School Program on Risk Regulation Research Paper No. 29. Available: http://scholarship.law.gwu.edu/cgi/viewcontent.cgi?article=1282 &context=faculty_publications.

248. National Science Foundation (2000). Nanotechnology definition. Available: http://www.nsf .gov/crssprgm/nano/reports/omb_nifty50.jsp.

249. M. Siegrist, "A causal model explaining the perception and acceptance of gene technology," *Journal of Applied Social Psychology*, vol. 29, no. 10, pp. 2093–2106, 1999.

250. L.J. Frewer, N. Gupta, S. George, A.R.H. Fischer, EL. Giles, and D. Coles, "Consumer attitudes towards nanotechnologies applied to food production," *Trends in Food Science & Technology*, vol. 40, no. 2, pp. 211–225, 2014.

4

The New Limeco Story: How One Produce Company Used Third-Party Food Safety Audit Scores to Improve Its Operation

Roy E. Costa

4.1 Food Safety in Modern Food Supply Operations

Food safety is a major concern of many food operations. When disease outbreaks occur, they create a crisis for the firms involved and may involve the media and agencies tasked with public health protection. The negative fallout can hurt a company, tarnish its image, and create a legal problem that may take years to resolve. Over the past 30 years especially, companies both large and small have focused on ensuring that the food they produce is safe as well as wholesome (food that not only does no harm but promotes healthy living).

> Unfortunately, the wholesome image that food companies have tried to cultivate has been marred by disease outbreaks.

For example, the Peanut Corporation of America (PCA), although only a minor supplier of peanut paste, managed to infect over 700 consumers of products that had utilized PCA's *Salmonella typhimurium*-contaminated peanut paste in 2008. Nine consumers died as a result. The outbreak resulted in the recall of over 3500 different products and highlighted dangerous practices that show the need for both governmental regulations and food safety systems [1].

Communication Practices in Engineering, Manufacturing, and Research for Food and Water Safety, First Edition.
Edited by David Wright.
© 2015 The Institute of Electrical and Electronics Engineers, Inc. Published 2015 by John Wiley & Sons, Inc.

PCA had detected *Salmonella* in its products many months prior to the outbreak, but it continued to send contaminated products out after "lab hunting." Lab hunting is the poor practice of sending food samples to numerous testing facilities until one lab does not detect contamination. That lab is then reported as the safety-certifying agent and previous contaminant-positive tests are suppressed. More than anything, PCA's tragic and criminal conduct reinforced the idea that firms should be held to a set of internal controls over the management of their food safety programs. Internal controls require trained supervisors and employees working under a set of safety standards tailored to the organization, including redundant safety controls. Internal controls also require transparency in food safety programs and conscientious and diligent oversight by upper management.

Additional wide-scale foodborne illness outbreaks (both before and after the PCA scandal) due to contaminated fruits and vegetables, potpies, "veggie" snacks, ground beef, peanut butter, and eggs led to the concept of food safety management systems (FSMS). The FSMS is a means of establishing the aforementioned internal controls. Such requirements for an organized approach to managing food safety have now been widely incorporated into the requirements imposed upon suppliers by the nation's largest retailers.

Retailers including restaurants and large chain groceries also suffer damage to their reputations when contaminated products reach consumers. Media coverage is usually extensive, and, like the food producers themselves, retailers also face the possibility of disastrous legal action and loss of business due to public perception.

Consumers demand action not only from the producers and retailers but form their governmental representatives. In 2011 President Obama introduced legislation designed to make the US food supply safer. *The Food Safety Modernization Act* [2] requires food companies to implement systems designed to stop contamination before it starts and allows food companies to issue mandatory product recalls if contamination is detected. He did so out of concern for public health and for his daughter, who enjoyed peanut butter sandwiches, a commodity recently involved in a foodborne illness outbreak, and, no doubt, for others around the country who enjoyed peanut butter as well.

The United States Food and Drug Administration (FDA) has jurisdiction over most foods produced in this country, while the USDA has primacy over meat. The FDA has new enforcement powers under the Food Safety Modernization Act and can now also force a recall of foods in addition to being able to inspect and regulate much of the nation's food supply. Many experts believe these new powers granted to the FDA will further support the safety of the U.S. food supply [2]. However, additional improvements are being actively encouraged by retail buying firms—those firms that purchase food products directly from the field for resale to consumers (grocery chains, restaurants).

As noted earlier, one of the important initiatives undertaken by food industries in the United States and abroad is the development of food safety management systems. These systems are comprised of standard operating procedures (SOPs) for all important production steps. SOPs provide for sanitary conditions, preventative maintenance, microbial testing, pest control, and personal health and hygienic standards, among others. A firm establishes the operations by describing who will do the task, what necessary equipment is required, the frequency of the task, and the specific location or operation involved. For example, an SOP for cleaning would prescribe a particular detergent, describe the proper method for applying that detergent to a particular piece of equipment or facility feature, schedule how often the cleaning is to be done, and the cleaning tools needed. Bolstering these efforts, FSMS also require training programs designed to provide employees with fundamental knowledge concerning their health, hygiene, and proper cleaning methods, as well as more involved processes that might include pasteurization, antimicrobial addition systems, traceability protocols, and Hazard Analysis Critical Control Point (HACCP) foodborne illness prevention programs. HAACP programs are designed to prevent food contaminations through chemical, biological, or physical elements and are unique in that they were designed to prevent contamination during production rather than as a system for inspecting finished products.

Today, training programs are often mandated by government agencies such as local and state health departments. For example, almost all states require managers of restaurants to receive food safety training and certification. In addition, both the FDA and USDA require their poultry and meat facilities to receive training in HACCP. However, many food producers and retailers are voluntarily implementing FSMS that go beyond governmental regulations. For example, sanitation training of hourly workers in restaurants has become a national trend, and many food production companies are choosing to hire sanitation managers in addition to creating their own training programs aimed at improving food safety.

Based on research conducted by the FDA [3], the presence of a certified sanitation manager in operations is associated with a reduction in the factors that lead to foodborne illness at the retail level. Certified sanitation managers, such as those certified through the National Registry of Food Safety Professionals, Prometric, or the National Restaurant Association, oversee the sanitation practices of a facility, process control procedures like cooking and storage temperature maintenance, and employee training.

In my experience, such training may not be effective in operations where standards are not in place, supervision is inadequate, or the operation fails to monitor and assess compliance with standards.

The food industry has seen a need, therefore, to create a framework for food safety efforts that ensures that standards are applied properly. My work as a food safety consultant (third-party auditor) requires me to ensure that standards are being met.

In the fresh produce industry, and in many other food sectors, the use of a third party to audit operations has become commonplace. Retailers use third-party audits to determine a company's conformance to a set of prescribed safety standards. Again, this development was a direct result of the foodborne illness scandals that had plagued food producers. Unwilling to be lumped in with producers as sloppy operators who did not care about their customers, retailers began demanding audits to ensure safety and quality. Third-party auditors provide an audit score that can be a powerful motivator for food producers, because the score is openly shared with retailers and becomes a benchmark for the reputation of a production company. Typical audit criteria prescribe a passing score of 75% as a minimum, but certain buyers (manufacturers, processors, and retailers) have much higher requirements, and expectations are for scores above 90% and preferably above 95%.

4.2 Safety Audits Cause Some Level of Controversy

Some controversy surrounds the issue of effectiveness of third-party audits and scores [4] as a means of preventing foodborne illness outbreaks. The audit score itself, being simply a snapshot of conditions at a particular time, may not always indicate the future risk of an operation. Some facilities have experienced outbreaks of foodborne illness after scoring well on third-party audits, causing some experts to question their validity.

Furthermore, since many firms supply food products to more than one retailer or manufacturer, a supplier may be asked to undergo several audits to satisfy the particular requirements of each buyer, which can be both costly and time consuming. So, there is continual debate about the redundancy of the current third-party audit system.

However, companies that develop FSMS and verify their performance through the use of audits find, in many cases, that their efforts improve their operation. The California Leafy Green Marketing Agreement (CALGMA) reported that there are three advantages to tracking audit citations among its members. Citations provide a strong foundation for training, provide mandatory corrective actions designed to reduce the number of repeat citations, and provide a basis for identifying problem trends over time. In its latest annual report, the CALGMA reports a 20% drop in citations for infractions of Good Agricultural Practices by third party auditors [5]. Reducing citations is important, because GAP infractions can lead to outbreaks.

The relationship between audits, scores, and performance is dependent upon the culture of the organization. My experience as a third-party auditor, which is based on hundreds of audits, has proven to me that when the culture of a food organization is driven by management commitment to the health of its consumers, food safety programs become highly effective in preventing foodborne illness. In such a culture, audits and scores become meaningful as measures of success within the organization.

In 2009, Frank Yiannas, vice president of Food Safety for Walmart, agreed [6], noting that consistent commitment by the management to safe operations is the best way to create a culture of safety. Food safety culture equates to what people actually do when producing and handling food. According to Yiannas, a firm builds a culture of food safety in several ways, but leading by management example is an essential part of the process. When the leaders of a firm exhibit a respect for food safety both in public and private it sends a clear message of the importance of food safety to everyone in the organization.

4.3 New Limeco's Journey to Safety

New Limeco is a medium-sized firm located in Homestead, Florida, and an example of one company that has experienced great success with food safety management. I have been working with the firm as an internal auditor for the past five years. New Limeco is an important provider of tropical fruits and vegetables, including avocados, boniato, carambola, and coconuts to South Florida markets. New Limeco grows avocados in South Florida and packs them at its home facility, while simultaneously storing and distributing many varieties of imported fruits and vegetables.

In 2008, the firm began the process to become food safety certified, responding to the requests of several of its customers. Most of the large buyers of produce in Florida now require that firms selling to major supermarkets have a verifiable food safety system. Those buyers are reacting to numerous outbreaks of foodborne illness due to *Salmonella, E coli* 0157:H7 and related strains. Such outbreaks can severely cripple the entire food supply chain because of recalls, product loss, and loss of consumer confidence.

For example, in 2008, *Salmonella* was found in Florida tomatoes and later in peppers. The tomato industry in Florida suffered huge losses by having to perform nationwide product recalls, with several growers going out of business even though the major vehicle turned out to be peppers coming from Mexico [7]. Such experiences have heightened concern, and many producers and retailers now feel that the best way to avoid these issues is to require certification of produce firms against recognized, benchmarked standards for food safety. Food safety certification, developed in response to such calamities, is a long process that requires development of science-based operating standards and process controls, training, record keeping, and qualifying through an independent evaluation performed by third-party auditing firms.

The Global Food Safety Initiative (GFSI) emerged as a leading nonprofit organization in developing certification standards for third-party audits. Founded in 2000, GFSI offers benchmarking standards for auditors and serves as a repository for food safety information. Private firms, such as Primus Labs, act as third-party auditors in implementing those standards. In New Limeco's case, my role was to act as consultant for Primus Labs in developing a certification program.

Circumstances required that the company develop a FSMS based on protocols provided by the major retailers through GFSI. The company responded initially by organizing itself around food safety and appointed a food safety manager with a strong regulatory standards background (prior experience in the nuclear energy industry), to lead the effort and prepare for its first audit. The manager developed FSMS that covered all operations, including purchasing, receiving, cooling, storing, grading, packing, and shipping of its lines of fresh produce.

Purchasing standards were perhaps the most difficult to implement, as the company sources hundreds of products from hundreds of suppliers in the United States, Central and South America, and the Caribbean. According to GFSI standards, all suppliers need food safety audits and New Limeco is required to have those approvals on file. While most of the other requirements were internal, New Limeco continues to work with external suppliers to finalize audits five years later.

Within its own organization, New Limeco, like many agricultural firms, faced the challenge of meeting hundreds of science-based standards. Most of the staff at the company is bilingual, with English as a second language, and there were no employees with a science background in 2008. The concepts of sanitation and the demands of the new programs that included a rigorous HACCP system, many technical procedures, and thorough record keeping were overwhelming at first. Because New Limeco is a small, family-operated company, there was also concern that the costs would be prohibitive.

In addition, the mostly Hispanic workforce seemed unprepared to deal with many of the new mandates. Employees were apprehensive that new duties required of them would be too difficult and time consuming (in addition to their already hectic work environment). The firm overcame these problems by slowly implementing many of the required changes, by supporting employees with a needed break area, and by listening to their concerns.

4.3.1 Implementing Changes

The changes first required that all employees be made aware of their new responsibilities under GFSI standards. The company hired a young production supervisor (in addition to the regulatory supervisor mentioned earlier) to help train employees and work with them in day-to-day operations. The new supervisor also had safety operations experience, so he was able to identify necessary food safety roles for employees and work them into the operation. The supervisor also developed procedures to ensure that cleaning (in particular) and maintenance of the facility were properly performed and documented. Finally, the supervisor developed a temperature-recording program for avocados to ensure that safe temperatures were maintained after the fruit arrived from the field. The company invested in automatic digital temperature recorders and tied those readings into a computerized system. Such efforts improved the safety of the product and reduced spoilage and waste.

Storage areas for other imported fruits and vegetable remained a problem initially, as the facility is basically an open-air operation, which is common for South Florida produce-packing operations. The firm developed systems to keep items covered, however, to minimize exposure of products to the environment. GFSI standards also require traceability systems, so the firm invested in ScoringAg$^©$ software as a means of creating labels and bar codes for its packed products. This sophisticated system allows for coding of products from the field upon intake and tracks them all the way through shipping so that, in the event of a product recall, the firm can find and remove affected items.

Throughout the first year, the food safety manager continued to implement changes to the operation. Through meetings and training sessions, personnel were educated as to how the new sanitation programs would work. To facilitate training, the firm constructed a special room in a previously unused space. Employees first needed training concerning new operational issues and, second, to familiarize them with their new responsibilities. The supervisor conducted the training using the new GFSI guidelines and prepared documents in Spanish as needed.

As the third-party auditing consultant, I made numerous visits to the site and initially spent several days at the facility, helping to troubleshoot operational issues involved with handling the firm's large and diverse product line. Most importantly, I assisted the management in conducting self-audits of FSMS, providing feedback to the food safety manager, who in turn shared the self-audit results with staff in the form of computerized GFSI reports with an analysis of findings and corrective actions that needed to be documented. The supervisor then coached employees on specific responsibilities to make sure that they made corrections effectively and in a timely manner.

4.3.2 Sanitation Issues

The most important aspects of the program involved personal hygiene issues, sanitation of the plant, pest control, microbial sampling programs, and monitoring of temperatures for sensitive products. The management supported the new programs by purchasing effective temperature-recording devices and a traceability labeling system, and by developing protocols for cleaning and personal hygiene of the staff.

Personal hygiene requirements started with training employees in the use of hand washing sinks, and proper sanitation of hands after using the restroom, but also at the beginning of work and after breaks. The firm created a large sign of employee best practices and posted that sign in a prominent location in the plant. The training and signage included employee health rules about reporting illness and not working while ill. As the New Limeco structure dates from the 1930s proper sanitation of the plant had its challenges. Areas that were neglected prior to the food safety program, such as floor finishing and walls, needed to be addressed. The firm repainted much of its operational areas and sealed off areas that were problems.

Bird and rodent control are also important sanitation issues for an open structure. The firm worked with its pest control service in developing an Integrated Pest Control Management Program with frequent inspection and maintenance of rodent traps. The firm installed spikes in bird-roosting areas and made other improvements to deter birds, such as prompt and ongoing removal of wastes from the operation and the closing and sealing of waste disposal areas.

Finally, the construction of a screened-in employee break area was another significant development. Prior to having a break, area employees took their lunch sitting on the concrete loading dock or on the steps to the shipping office. There was only a very small refrigerator and no microwave. During inclement weather the employees were forced to eat lunch in production areas. GFSI does not permit such practices for several reasons, most importantly to avoid cross contamination of equipment and the environment. Furthermore, packing fruit and vegetables is physically demanding work and the comfort of a screened break area with tables and chairs, sinks and refrigerators was greatly appreciated by the staff and helped to solidify their commitment to the safety program, sending a clear message to employees that they were an important part of the food safety program.

4.3.3 Gradual Safety Improvement

Although my first preliminary audit was less than *Excellent* (below 90%), as the scores on self-audits began to rise, employees responded to each new score enthusiastically. New Limeco began to experience several benefits as employee motivation began to increase. The facility was much cleaner, better organized, and more efficient.

When the time came for me to officially evaluate the firm, employees rallied together to make final improvements. The first official audit resulted in a *Superior* rating for the firm (above 95%). The impact of this score had a measurable positive effect on employees. The firm celebrated, and the employees felt a deep sense of accomplishment.

New Limeco's food safety program gained momentum from this first audit, and the firm has sustained its performance with *Excellent* and *Superior* scores over the course of the last four years. New Limeco continues to make improvements, most recently developing and implementing an advanced labeling and product traceability and recall system. Consultants now visit New Limeco only twice per year to maintain a high level of conformance, which the firm has done even with increasingly intensive third-party standards, such as those required under GFSI. The employees await each external third-party audit with high expectations, and the firm communicates its scores to the public through its website and promotional material.

In addition to the success of its food safety program, in 2010 New Limeco also won a prestigious Sharp Award, presented by the University of South Florida for its compliance with requirements of the U. S. Occupational Safety and Health Administration (OSHA).

The award commends Florida's employers and employees in all industries who proactively and routinely engage in job safety. To qualify for this honor, New Limeco was evaluated on administration and oversight of safety management programs, safety and health management practices, safety training and instruction, employee safety rules, and internal safety inspection routines for new and existing substances, processes, and equipment.

While some food firms might view third-party audits as an intrusion, firms with a strong food safety culture, such as New Limeco, benefit greatly from the process of external evaluation and scoring.

References

1. M.R. Taylor, "Will the Food Safety Modernization Act help prevent outbreaks of foodborne illness?" *New England Journal of Medicine*, vol. 365, no. 9, pp. e18 (1–3), 2011.

2. U.S. House, 112th Congress. H.R. 2715, Food Safety Modernization Act, January 15, 2011. Available: http://thomas.loc.gov/cgi-bin/bdquery/z?d112:h.r.2715.

3. FDA National Retail Food Team (2009). FDA Report on the Occurrence of Foodborne Illness Risk Factors in Selected Institutional Foodservice, Restaurant and Retail Food Store Facility Types. U.S. Food and Drug Administration, Washington, DC. Available: http://www.fda.gov/Food/GuidanceRegulation/RetailFoodProtection/FoodborneIllnessRiskFactorReduction/ucm224321.htm.

4. D.A. Powell et al., "Audits and inspections are never enough: A critique to enhance food safety," *Food Control*, vol. 30, no.2, pp. 686–691, 2013.

5. M. Villeneva, "Three reasons why audit citations are a good thing," LGMA News, November 12, 2012 [Online]. Available: http://lgma.ca.gov/node/153.

6. R. Yiannas, *Food Safety Culture: Creating a Behavior-Based Food Safety Management System.* Springer: New York, 2009.

7. E. Taylor, J. Kastner, and D. Renter, "Challenges involved in the salmonella saintpaul outbreak and lessons learned," *Journal of Public Health Management & Practice*, vol. 16, no. 3, pp. 221–231, 2010.

5

Communication Practices by Way of Permits and Policy: Do Environmental Regulations Promote Sustainability in the Real World?

Becca Cammack

5.1 Communication in the Modern Environmental Movement

Communication has played a key role in the development and evolution of the postmodern environmental movement since its emergence in the late 1960s. The consequences related to our clear misuse of the environment were originally unveiled to a seemingly ignorant populace by way of protests and by radical environmental groups. Today, the sense of urgency surrounding environmental degradation and the human contribution to this has only escalated, with environmentalism no longer viewed as a radical concept. In fact, most of today's civilized world has grown to recognize the importance of our environment and the role that humans play in either its ongoing degradation or seeming unlikely salvation. There is no doubt that this shift was brought about by effective means of communication, communication facilitated by science, scholars, and the environment itself.

The early environmental movement emerged during the Industrial Revolution [1] in response to the widespread pollution of our air, streams, and rivers. This young and volatile movement represented our earliest attempts to organize and communicate the need for environmental accountability. Over the decades, however, grassroots efforts of the movement's infancy including marches and protests have given way to a more mature network of legislation and policy. In clear opposition to the free-spirited liberal

Communication Practices in Engineering, Manufacturing, and Research for Food and Water Safety, First Edition. Edited by David Wright.
© 2015 The Institute of Electrical and Electronics Engineers, Inc. Published 2015 by John Wiley & Sons, Inc.

attitude associated with environmentalism, regulations now mandate the ways in which we use our water, air, and natural resources. Administered by a complicated network of regulatory agencies, environmental regulations mark the maturation and legitimization of a movement once considered radical. These regulations also provide a tangible means by which sustainability—the ability of human beings to sustain our environment while maintaining long-term ecological balance—can be both communicated and measured.

This project involves examining the communication practices of environmental regulations. Specifically, I examined the ability of these regulations to convey the importance of incorporating sustainability into our thoughts and lives. A vital question to be asked is whether a connection exists between environmental regulations and our perception of sustainability. Do environmental regulations actually encourage sustainability at a cultural level? Have we grown more sustainable in our awareness, thoughts, and behaviors as a result of being thrust into the mind-set of compliance?

While these questions are expansive, and beyond the scope of a single study, this project focused on a single regulation and its direct impact on members of the professional community. The California State Water Resources Control Board's (SWRCB) Construction General Storm Water Permit (CGP) is a construction permit issued to developers and builders in California when their development projects or construction activities meet or exceed 1.0 acre of soil disturbance. The CGP (and similar construction storm water permits in other states) stems from federal regulations, such as the Clean Water Act (CWA), which were enacted to regulate and manage the release of construction-related pollutants into storm water.

To study the persuasiveness of the CGP, rhetorical and textual analyses were performed to determine how, or if, this permit communicates the concept of "sustainability." Additionally, a small case study was conducted involving a group of environmental professionals with active experience working within the restrictions of the CGP. The ultimate goal of this research was to determine how, or if, these individuals have incorporated sustainability into their consciousness, decisions, and daily behaviors as a result of working within the constraints of environmental regulations. Have their values or views on sustainability changed as a result of their exposure to the permit? Have they reconsidered how their work or personal actions can impact the environment? Or, conversely, does the CGP simply prescribe a set of rules and constraints to people who have no choice but to conform?

5.2 Background

Environmental policy and regulations are enacted primarily in the form of permits issued to companies or individuals by regulatory agencies [2]. Once issued, these permits legally allow specific activities—ranging from short-term development projects to ongoing commercial operations—to be carried out. Over the last several decades, environmental permits and the regulations they stem from have played an increasingly prevalent role in most commercial operations, with few industries left untouched by the cost and effort involved in maintaining environmental compliance.

5.2.1   Who Is on the Receiving End of Environmental Regulation?

While enacted to restrict how a certain segment of the population is able to use and impact its local resources, many environmental regulations actually have very little jurisdiction at the residential level. Despite the impact that a single individual or family can have when choosing to live a sustainable lifestyle (or conversely, when defaulting to a wasteful one), many regulations and their related permits do not target the residential sector. Instead, the majority of regulations are written for and enacted within a target group considered to be more manageable from a regulatory perspective (and far more lucrative from a financial one): the commercial industry [3]. The industrial and commercial sectors, being the drivers behind large-scale development and expensive industrial operations, are also the typical culprits behind pipes discharging oily wastes into rivers, smoke stacks spewing clouds of gas into the atmosphere, and hazardous wastes. In short, the commercial/industrial sectors represent a highly visible source of environmental degradation. As such, these sectors are also the recipient of hundreds of environmental regulations to have been enacted over the last 50 years. As development and industry become increasingly more regulated, a subsequent overlap into the planning and engineering fields has resulted in requirements and restrictions over how projects are designed, built, and maintained. Projects are now analyzed early in the planning stages to determine potential impacts to resources such as air quality, water quality, local species, and habitat. The resulting project-specific impacts dictate the ways that projects can be engineered.

The effects, both positive and negative, of this relatively young branch of regulation over the United States' commercial industry has been a subject of debate since the 1980s—the decade when big business and industry first began to feel the sting of regulatory compliance. These debates were highlighted to the public in several journal articles, such as *The Challenge of Going Green*, published in the Harvard Business Review in 1994. This article illustrates the debate among scholars and experts as they argue over the impacts of environmental regulations on US business. While some experts have remained optimistic that environmental regulations have the potential to actually bring about cost savings as wastes are reduced and operations streamlined, many others have argued that the cost of doing business around a barrage of environmental permits presents a dangerous strain to US business [4]. In either case, the myriad of environmental regulations has only continued to intensify, and today nearly every segment of US business has been forced to adapt to the rising cost of compliance.

5.2.2   What Are the Effects of Construction and Storm Water on the Environment?

Storm water runoff flows into conveyance systems through curbs and gutters, where it is then typically routed to the nearest body of water. In coastal areas, conveyance systems empty into rivers and streams leading to the ocean or directly into the ocean itself. This connection between storm water conveyance systems and receiving water bodies also provides an ideal migratory pathway for pollutants picked up by storm water migrating through urban areas [5]. Storm water falling onto streets and parking lots picks up oil and debris, while storm water falling onto construction sites and industrial work areas picks up chemicals and loose soil (sediment). The recognition of these different urbanized settings and their potential to impact storm water runoff led to the enactment of regulations, such as the CGP, to address this. The CGP, more formally known as California's SWRCB Division of Water Quality Order 2009-0009-DWQ, is a regulatory permit issued to construction projects having at least 1.0 acre of soil disturbance. Permitted projects can lawfully proceed given they meet very specific conditions resulting in the proper maintenance of disturbed soil and other pollutants onsite, thus minimizing the impact to local water quality by preventing the migration of pollutants to storm drains or other waterways [6]. In short, the CGP is intended to help construction crews manage the loose sediment and other pollutants produced on their job sites.

The SWRCB CGP is issued and overseen by California's State Water Resources Control Board. As of 2011, approximately 60,000 California construction projects (in various levels of execution or completion) had obtained coverage under the CGP statewide [6]. This requirement, however, is not restricted to California alone. The CGP is based on the EPA's own construction general permit for storm water—a permit providing similar restrictions and overarching requirements for construction activities at the federal level. The EPA's regulatory jurisdiction under this permit is nationwide and applies to any state or municipality not having its own local regulations (such as California's CGP). Thus, it can be concluded that most of the qualifying construction projects and activities (resulting in one or more acre of soil disturbance) throughout the United States are subjected to some form of regulatory oversight pertaining to soil disturbance and water quality, whether it be at the federal level (through the EPA) or by a more local form of government (such as California's SWRCB). Municipal storm water permits present a higher level of specificity, being enacted at the city or county level and thus taking regulation and enforcement one step further. These permits and their enforcing agencies emphasize sediment control, as based on scientific (primarily EPA) data finding that sediment is one of the most prominent pollutants now found in the waters bodies downstream from construction sites. Studies also show that the sediment found in these water bodies is primarily shed from construction sites and other nonstabilized surfaces located upstream.

The CGP provides requirements and restrictions surrounding how soil-disturbing activities are to be carried out, with the overall goal being the minimization of sediment and pollutants discharged from project work areas. When a construction project triggers coverage under the CGP, permitees are required to submit notifications and fees to the SWRCB. The CGP also requires that qualifying projects document site conditions and demonstrate that work activities meet permit requirements. Permitted projects are required to develop project-specific plans detailing the means by which storm water pollution will be prevented, to monitor onsite construction activities, to document that storm water flowing off of the construction site is pollutant-free, and to stabilize all areas of disturbance at the end of the project [6]. Conformance with these permit requirements often comes at a high price: cumulative costs for implementing each of the CGP's requirements throughout a project's life cycle can add thousands (and in some cases, hundreds of thousands) of US dollars to project costs. Many of California's construction companies and land developers have struggled to remain in compliance with the seemingly exhaustive terms of the CGP, especially as it grows more stringent with each permit revision. Additionally, some members of California's construction community continue to question the effectiveness of the CGP in actually protecting or improving water quality. Critics of the permit have questioned what, or if, it does to help or protect the environment, especially in the face of government budget cuts, which have limited the resources available for consistent enforcement.

The long history and controversy surrounding the CGP have led us to ask certain questions:

- What exactly is being communicated in the CGP?
- Does the CGP encourage people to look at the world and their impact from a different light?
- Above all, does the CGP communicate "sustainability"?

5.3 Studying Groundwater Regulation

These research questions were addressed in two separate efforts. First, a textual analysis was conducted on the CGP to identify the presence and use of persuasive elements. This analysis allowed the CGP to be examined in how it communicates to and ultimately persuades its users. It was also examined for the presence of underlying themes and rhetorical strategies that could be viewed as persuasive in encouraging CGP users to accept a sense of environmental responsibility, and ultimately to adopt more sustainable behaviors. In a separate effort, a small case study was conducted involving a specialized group of environmental professionals with experience in working on projects covered by the CGP. These professionals have worked on numerous long-term, high-visibility projects subjected to CGP coverage and have watched it evolve through several revisions. Thus, these individuals are appropriate candidates to discuss the CGP and attest to its influence. The case study (though limited in the number of participants) allowed an evaluation to be made on whether or not the CGP actually communicates in terms of promoting real-life sustainability.

Comprehensively, these two efforts allowed an initial analysis of the means of communication utilized in the CGP, and a more comprehensive assessment as to whether this means of communication is effective. Finally, these efforts allowed conclusions to be drawn regarding the message conveyed by the CGP and if this message holds power in practice.

5.3.1 Textual Analysis

The means of communication and primary messages conveyed by the CGP were studied by way of a textual analysis. This analysis involved studying the structure, organization, and content of the CGP to identify how, or if, it was constructed rhetorically, and if it is, in fact, a persuasive text. Being a regulatory permit, the CGP provides requirements and restrictions for projects meeting the criteria for coverage and is accompanied by a separate attachment: the CGP Fact Sheet. This Fact Sheet is provided as a supplement to the CGP, and was prepared to justify the changes and revisions imposed when the permit was updated in July 2010. For the purposes of this textual analysis, the Fact Sheet and CGP were addressed separately.

5.3.2 Case Study

Specific questions pertaining to the ways in which the CGP communicates and the messages it conveys were posed to members of an environmental team having specific experience in working with CGP-covered projects. These study participants ($n = 4$) were asked to talk about what the permit ultimately communicated in terms of its persuasive elements. Likewise, the participants' responses and communications were evaluated based on their responses to the questions.

Survey questions were distributed to the case study participants, who were asked to provide detailed responses on the influence of the permit on one's own mind-sets and behaviors, in addition to their observations of its influence on others. We summarize the responses in Section 5.4.

5.4 Results of My Investigation

The CGP Fact Sheet was prepared by the SWRCB to explain and justify the numerous revisions and new requirements proposed during the permit's most recent revision in 2010. As noted earlier, many of the proposed revisions and new requirements were a controversial topic of discussion among members of California's construction industry in the months (and even years) leading up to the 2010 revision. Many members of California's community viewed the proposed requirements as extreme, costly, and, at times, even irrelevant in terms of protecting water quality. As I witnessed in e-mail/online forums, it was also widely discussed among members of the construction industry that the proposed revisions would cause the costs of development in California to escalate even further, and that compliance with the related permit requirements would require more time, energy, and accountability on the part of the builders. The SWRCB made an effort to address these comments and concerns with the Fact Sheet, and did this by way of two main discussions:

1. Background
2. Rationale

The organization of the Fact Sheet into these two sections, Background and Rationale, demonstrates the importance of simply and logically presenting the need for the permit's new requirements to its users. Quite simply, these discussions first provide a history of the CGP and pertinent water quality issues (Background), followed by a rationale or justification as to why the more stringent changes should be adopted in the new revision of the permit (Rationale). The two sections and their subsections are supplemented with tables and figures, each of which presents scientific data used to further validate the claims made in the Fact Sheet.

5.4.1   The CGP Fact Sheet Background Section

The "History" subsection of the Background Section provides a brief background of the water quality laws supporting the regulation of storm water, dating back to the adoption of the Clean Water Act (CWA) in 1972, when it was enacted to address industrial pollutant sources. Moreover, it discusses the evolution of the CWA throughout the 1970s and early 1980s, at which time it was amended to incorporate storm water within its jurisdiction of regulation. This leads to a discussion of EPA regulations from the 1990s, in which the concept of focusing on construction sites based on their association with soil disturbance was introduced. Finally, the evolution of these construction permits throughout the 1990s and 2000s, leading to the revisions and adoption of the existing CGP, is discussed. This section discussing the history of storm water law is important rhetorically as it lays the groundwork for current initiatives to control storm water pollution.

"Legal Challenges and Court Decisions" summarizes important court cases and legal decisions taking place following the original adoption of the CWA, ultimately leading to the recognition of storm water as a source for pollution. The significance of this recognition is the ultimate incorporation of storm water into the network of regulatory oversight over water quality issues. While a number of these cases initially challenged the regulation of storm water discharges, the resulting judgments ultimately supported the ongoing evolution of the CWA and continued to bring storm water into a central focus of water quality law. This contributes to the rhetorical framework of the Fact Sheet by again reinforcing the significance of storm water and storm water pollution from an overarching regulatory perspective, which supports current and smaller-scale efforts to regulate storm water pollution (such as the CGP).

The next subsection, entitled "Blue Ribbon Panel of Experts and Feasibility of Numeric Effluent Limitations," provides a context into which the CGP's proposed revisions were investigated. This is an important rhetorical move as it identifies the fact that the requirements and restrictions proposed in the 2010 CGP revision were investigated by a team of experts. Prior to the adoption of the revised CGP, the SWRCB tasked this panel with analyzing the feasibility of the proposed changes, specifically those that would significantly impact the existing approach for construction projects. The Blue Ribbon panel investigated the potential impacts of the proposed requirements and examined how related compliance actions might be carried out. The panel was also tasked with investigating how permit holders would go about complying with the numeric limitations should they be incorporated into the CGP. The simple fact that the SWRCB took major steps to investigate the feasibility of the changes they were proposing anticipates and attempts to correct their critics' concerns surrounding the revisions.

The "Summary of Panel Findings on Construction Activities" section summarizes the findings of the "Blue Ribbon Panel," which ultimately recommended that numeric effluent limits and other proposed changes be incorporated into the CGP. This section is comprised of direct quotations from panel members, in which they state that the proposed changes are not only feasible but also advisable based on similar adoptions in neighboring Pacific states. They also state the general belief that monitoring for specific pollutants will be beneficial to the environment. The direct quotations from the panel lend to the validity of their findings, as the findings appear to be official: "It is the consensus of the Panel that active treatment technologies make Numeric Limits technically feasible for…larger construction sites" [7, p. 5]. The quotes also place ownership of findings onto the panel versus the SWRCB, which potentially diffuses criticism. This also further supports the rhetorical framework of the Background section of the Fact Sheet; through discussing the Blue Ribbon panel and its efforts, the CGP and its content are further reinforced and legitimized.

The final subsection, "How the Panel's Findings Are Used in This General Permit/ Summary of Significant Changes," provides a brief statement regarding the findings of the blue ribbon panel to recommend that the proposed changes be implemented, and how these findings essentially resulted in their incorporation into the CGP. Essentially, the panel had identified gaps in the pre-existing version of the CGP, and those gaps would be addressed by each of the proposed revisions. This concise conclusion serves to bring a sense of closure both to the Background section as well as the debates surrounding the revisions. With the regulation of storm water now having been discussed from the inception of water quality law through the investigation and recommendations of subject-matter experts, the relevance of the proposed changes has been established.

5.4.2 The CGP Rationale Section

The second section of the Fact Sheet, "Rationale," summarizes the changes adopted in a concise list, item by item: the approach of the permit, the types of construction activities that are covered, the means to obtain permit coverage, and the requirements to stay in compliance with the permit requirements. Significant changes, particularly those pertaining to the numerical effluent limits or other major impacts to project design and execution practices, are explained in detail. Accompanying figures and tables are used to illustrate the scientific data backing the changes, further justifying them. The "no-nonsense" presentation of the new requirements and incorporation of data, particularly the graphs and figures that serve to visually depict the processes where raw data could not, further supplement the framework set by the Background section's lengthy narrative. The use of data has the effect of "sealing the deal" from a communications standpoint, silencing all objections with the simple presentation of logic.

The second part of the rhetorical analysis was performed on the actual CGP.

5.4.3 Construction General Permit (CGP)

Unlike the fact sheet, the CGP does not incorporate attempts to explain or justify its restrictions and requirements. This is to be expected, as the CGP has a completely different function and purpose than the Fact Sheet. As a regulation, the CGP prescribes rules, restrictions, and requirements that are presented logically and systematically, section-by-section. Similar to the Fact Sheet, however, the CGP opens its section on General Findings by discussing the Clean Water Act (CWA) and its prohibitions, and how storm water became a part of this federal water quality law. This opening reference to the CWA serves to link the CGP to the long-standing EPA and CWA, thus enhancing the CGP's credibility and weakening potential objections to its content. References to the CWA and EPA give the CGP the appearance of being more robust, grounded within a network of long-standing federal law and regulation. Among these are references to decisions against the EPA's requests to exempt certain industries and activities from storm water monitoring, such as the Ninth Circuit Court of Appeals 2008 NRDC versus EPA ruling that oil and gas construction activities were subjected to federal permitting requirements.

Even with its grounding in technical and regulatory terminology, however, the CGP still makes use of footnotes to provide definitions and explain its revisions and new requirements (similar to the justifications presented in the Fact Sheet). These footnotes provide a means for explaining the basis behind the new requirements. Other strong communicative tools used in the CGP involve the use of action words such as "require," "limit," and "prohibit." By clearly identifying permit requirements, limitations, and prohibitions, permitees are clearly informed of what they can and cannot do. In addition, the CGP language is recognizable as a standard operating procedure in that it clearly lays out directives by informing permitees how they "shall" fulfill permit requirements. These different tools help to sell the legitimacy of the CGP. The organization and presentation of the CGP lends to an official and professional feel of the document, thus leading to a wider acceptance of its content and a higher tendency to follow its terms.

5.4.4 A Targeted Case Study of CGP

To study the efficacy of the CGP in the real world, I conducted a targeted case study with four environmental professionals who work within California's construction community, and who have some degree of experience or expertise in working with the CGP and its requirements. These individuals serve as project managers and subject matter experts within project teams on complex projects designed, carried out, and completed around environmental regulations and restrictions. These large projects often trigger coverage under multiple permits, with the CGP being only one of the many environmental permits issued. The individuals targeted in this case study support projects from the planning phase, throughout all active phases of construction and development, and ultimately through permit closure efforts (which can often involve as much documentation and effort as the projects themselves). They also provide support and guidance to project players at all levels, ranging from construction personnel working on job sites and managing permits in the field all the way to corporate executives and high level managers who may never set foot on an actual job site. This in-depth involvement at all levels in the project provides these case study participants with the opportunity to see environmental permits such as the CGP enacted at all project levels. As such, these individuals are suitable to discuss the CGP in terms of both personal and professional influence.

Once the case study participants were identified and approached to participate in this study, case study questions were developed and distributed to them via e-mail. The questionnaire consisted of 13 questions posed in essay format that asked participants to discuss their basic knowledge of the CGP and then talk in more depth about their perceptions of the CGP's messages and if its communication could be tied to sustainability. The responses are summarized here.

1. **Has working with the CGP influenced your thoughts/attitudes towards the environment and sustainability?**

   Three of the study participants confirmed that, to some degree, the CGP had influenced them and their activities.

   The case study participants consisted of environmental project managers working alongside myself for a utility company in Southern California. We all worked together on numerous projects, and our jobs were to manage environmental permits throughout the life cycle of these projects (in the process discussed previously in this chapter). When embarking on this academic research project, I approached each individual to ask them about participating in this study. Following their consent I distributed the questionnaires to them via e-mail and then summarized their responses in this chapter. While each of my case study participants played a different role and had a different level of experience in working directly with the CGP (as opposed to other environmental permits, for example), they all performed daily project work impacted by the CGP and possess what I would consider to be a unique level of insight into this permit's requirements restrictions and costs.

   - Participant 1 observed that she now sees issues with water quality everywhere she goes. She talked of the evidence she observed of damage to trails and plants while out hiking in areas where, presumably, the other hikers would be also be nature lovers and somewhat concerned with sustainability and environmental quality. However, the degradation (such as littering) that she has witnessed to the resources in these areas has discouraged her. "These are the most likely people who support environmental initiatives. I wonder if they don't care or don't understand. If they (being nature lovers, hikers, etc.) are not going to follow the rules, is there any hope? Not sure if sustainability is possible but I do think we (humans) need to try to work towards sustainability."
   - Participant 2 stated that while he was already conscious of environmental issues as a member of the professional environmental field, since working with the CGP he had become more aware of water quality in general.
   - Participant 4 stated that, while also already being a member of the professional environmental community and thus aware of environmental issues, the CGP had advanced his general awareness surrounding storm water runoff and pollutants. "I have a greater appreciation for the widespread concerns about storm water runoff/pollution and how that may impact the environment. The need to protect the environment from these sources of pollution is a key ingredient for a company that is trying to look at the sustainability of maintaining its system within the community that it serves."

Only one of the participants, Participant 3, declined that the CGP had led her to incorporate new habits or behaviors into her life; however, she acknowledged that as an environmental professional with many years of experience in her field, she already felt that she was aware of related issues.

2. **Are there any new activities or habits (geared towards sustainability) incorporated into your personal life or behaviors as a result of your work with the CGP?**

   • Participant 1 reported that there were several activities specific to water quality that she had adopted as a result of her work with the CGP. While doing work in her yard at home, for example, she installed protective measures designed to retain soil and prevent it from leaving her yard and entering storm drains (with a similar type of protective measure prescribed by the CGP and commonly used on construction sites). She also had begun to avoid washing her car in the street, another activity that can result in the discharge of pollutants to the storm drain. In addition, she discussed making conscious efforts to stay on trails while hiking.

   • Participant 2 stated that he had not incorporated new behaviors into his life as a result of the CGP; however, he stated that as an existing environmental professional he was already aware of the potential impacts of his behavior and had already made conscious efforts to limit his ecological footprint.

   • Participants 3 and 4 stated that they had not incorporated new behaviors into their lives but did not provide an additional explanation as to why (or why not).

3. **Are there any prior routines or habits (perhaps not geared towards sustainability) that you have changed as a result of your work with the CGP?**

   • Participants 1 and 2 observed that there are activities they previously participated in, including driving off-road and washing off paint brushes in the sink, which they had eliminated as a result of their work with the CGP.

   • Participants 3 and 4 stated that they had not changed any of their habits, but they did not provide an explanation as to why (or why not).

4. **Have you witnessed a change in your attitude in working with the CGP since it was first introduced?**

   • Participant 1 observed that many of the permit requirements prescribed in the CGP are perhaps too stringent and, ultimately, are extraneous when it comes to protecting water quality. She acknowledged the need to implement best management practices for the protection of water quality and that some type of regulatory driver is needed to ensure that this happens; however, she also discussed the extensive requirements of the CGP in terms of exceeding what is really needed to protect water quality. "I think there are some really over the top requirements in the CGP that I do not think do anything to help water quality. I don't think it was thought out how it would be implemented. I do think there is a need for BMP's and they need to have a legal driver or they would not be followed but I think the CGP is cumbersome and many of the requirements do nothing for water quality." Ultimately, Participant 1 discussed the CGP's lack of consideration with regard to other types of environmental issues and how it could be a more effective permit if a multidiscipline approach was taken. "As a biologist, it really bothers me that the environmental agencies are so siloed. I would like to see some biology requirements in the re-vegetation requirements in the permit."

- Participant 2 observed that he was initially skeptical of the CGP; however, he has since recognized the permit's worth in providing clearer guidelines and rules for workers to follow. "I had a hard time swallowing the permit at first given the more formal reporting requirements etc. Having [now] had a chance to work with it for a while, I am not as critical of it and realize it creates more clear guidance and helps create more clear standards."

- Participant 3 acknowledged that the controversial revisions and stringent changes to the CGP were positive as they helped to create a better-rounded program.

- Participant 4 acknowledged the importance of having protective measures and written plans in place to protect water quality; however, he voiced frustrations with the administrative work involved with maintaining compliance with the CGP. He acknowledged a growing appreciation for those who work to maintain permit compliance. "The crews look at the job site and they now notice when the [materials] don't seem to be installed correctly, and for some road grading work there was some surprise that after a season of heavy rains, properly installed [materials] actually DO WORK."

5. **Have you felt compelled to communicate to others on environmental responsibility and sustainability as a result of your work with the CGP?**

- Participant 1 acknowledged that only in a very limited amount has the CGP driven her to discuss water quality with others. If she is out driving around with friends and observes something specific to water quality, she might point out the issue.

- Participant 2 observed that he already felt a responsibility towards educating others on environmental issues; however, he does not relate this sense of responsibility specifically to his work with the CGP.

- Participants 3 and 4 declined that they are driven to communicate with others based on the CGP, but they did not provide any explanation for this.

6. **Have you witnessed a change in attitude or behavior of others on a jobsite as a result of work with the CGP?**

- Participant 1 observed that she has seen a general change or acceptance of the CGP in construction managers on jobsites; however, she also noted her skepticism in this change being due to genuine concern for the environment. "I think the [construction managers] are trying to understand and follow the CGP. Not sure if there is a change in attitude of if they are just doing what they are required to." She notes that while some workers do seem to express a concern for the environment and more global issues, many others are simply more likely concerned with following the rules to save their jobs. "I think there are some [construction managers] who really want to try and others who only do it while you're looking."

- Participant 2 noted that he has witnessed an improvement in workers' attitudes toward the CGP and its requirements when he explains what the CGP is trying to accomplish. "Some of the construction personnel I work with have had slightly 'frustrated' attitudes with some of the additional obligations we have as a company. Explaining the permit helps them understand where the requirements come from and help soften their attitudes."

- Participant 3 declined seeing a change in attitude or behavior, noting, however, that she is also not present at jobsites very often to see such changes.

- Participant 4 acknowledged that crew members now observe when protection methods or work practices are not being used correctly and acknowledged that when used correctly they do, in fact, work.

7. **From your perspective, do you feel that the CGP teaches sustainability and environmental responsibility?**

- Participant 1 stated that, no, she does not feel the CGP teaches sustainability; it merely prescribes requirements and restrictions. "Not really. I don't feel it teaches anything. It has a lot of requirements but I don't see how it is teaching. I guess there is training but it just tells people how to stay in compliance not how to be sustainable."

- Participant 2 stated no, as the CGP does nothing to actually limit development or preserve natural resources. "I do not think that it does because in no real way does it slow/limit development. "

- Participant 3 was unable to answer the question as she had only limited experience with working directly with the CGP.

- Participant 4 expressed doubt, acknowledging that the day-to-day job site activities may not necessarily help someone come to an understanding of how these requirements fit into the larger concept of sustainability is not possible for many people. "For many, the leap from reducing, managing, controlling stormwater runoff to protect the downstream environment to assisting in the sustainability of the local environment is too large of a leap to make."

8. **How might the CGP be revised to better promote sustainable behaviors?**

- Participant 1 does not feel that the CGP teaches more sustainable behaviors, and that it is simply a guide to performing construction activities. "I guess I don't know if there is a goal in the CGP to promote sustainability. I feel that it is strictly a permit to guide construction activities and doesn't really do much to promote sustainability." In order to learn sustainability, permit users would have to educate themselves by some other means outside of the permit. "If you are interested in sustainability and you work with the permit, you would become aware or educated in ways to change your actions to be more sustainable."

- Participant 2 also feels that there should be more collaboration between the CGP and other types of environmental regulatory permits. "There is not enough overlap and collaboration between CGP and other agencies. I feel this is a problem with agencies in general, not just with the CGP. For example, we have to get closure on sites that may also be sensitive cultural resource areas where we need to limit or avoid ground disturbance." He discussed the importance of taking natural resources and other types of environmental issues into account in order for permit users to learn about sustainability.

- Participant 3 was unable to answer the question as she feels that storm water is not her area of expertise.

- Participant 4 feels that the CGP should focus more on the areas where storm water runoff ends up in an effort to control/manage pollution, as opposed to focusing all efforts on managing materials on jobsites. "[There should be] greater focus on looking at where storm water runoff might have an actual impact on the environment and not be as concerned with making sure that everything is in place in those areas less likely to be impacted."

5.5 Discussion of Study Results

This project involved examining the concept of sustainability as it is communicated by environmental regulations. Specifically, this project focused on identifying what has been communicated by California's CGP. In brief, I wanted to know if, and how, this environmental regulation communicates with its users and if this communication ultimately causes them to incorporate sustainability into their thoughts, lifestyle choices, and behavioral patterns. Do the people who work underneath these regulations begin to view the world, and how they impact it, differently? Do they incorporate new and more sustainable actions into their lives as a result of being thrust into the mindset of regulatory compliance? Do they avoid choices and actions with the potential to harm the environment because of their exposure to these regulations? Or, conversely, do environmental regulations simply prescribe restrictions and requirements that mandate how work is carried out, without having any significant educational or persuasive sway?

My textual analysis identified the presence and use of rhetoric primarily in the CGP Fact Sheet, a document that was prepared by the SWRCB to accompany the CGP when it was revised and to justify the (frequently more stringent) revisions. In contrast to the CGP itself, the Fact Sheet provides an in-depth background discussion on storm water laws and their relevance to long-standing and widely recognized environmental issues. The Fact Sheet originates with a discussion of the Clean Water Act in the 1970s and goes on to narrate its evolution as the significance of storm water pollution was gradually recognized and then incorporated into water quality law. This association with the CWA and federal court decisions proves to be effective rhetorically as it provides an official grounding for the CGP. And, conceptually, with this association, the CGP transitions from a set of frustrating rules administered by local enforcement agencies (as perceived by some) into an important contributor toward a global initiative: the protection of our water quality.

The Fact Sheet also makes use of supporting scientific data in the form of graphs and tables, thus further validating the new requirements and restrictions incorporated into the 2010 revision of the CGP. With California's construction community viewing many of the proposed changes as extreme, costly, and even somewhat irrelevant to the overall protection of water quality, the Fact Sheet's use and presentation of scientific data again serves to link the CGP to larger environmental issues, the significance of which are difficult to refute. When presented with visual representations of processes such as erosion and sediment loss (as opposed to merely discussing them), the proposed changes are perceived more as a necessary means to an end (as opposed to a senseless and costly network of rules). In short, the Fact Sheet's use of scientific data, the connection between the CGP and long-established federal water quality laws, and the logical connection between strict work practices and environmental conservation are sound communication methods in which permit users are persuaded to accept the CGP.

Results of the case study were also illuminating. Generally, participants either incorporated a new practice into their lives or eliminated an old one as a result of their exposure to the CGP. While two participants actually disagreed that the CGP had influenced them in this way, they also stated that (as long-time members of the professional environmental community) they had already incorporated sustainable behaviors into their lives as a result of their professional work and exposure to similar environmental permits. Additionally, over time all participants witnessed a change in perception—both in themselves and others—towards the CGP itself. While there may have been an initial sense of resistance to the new requirements and revisions, a gradual sense of acceptance and understanding seemed to generally take place. These responses would indicate that the CGP does, in fact, effectively communicate the need for sustainability and acceptance of regulatory-driven behavior.

However, when asked if they felt that the CGP directly teaches or compels one to live sustainably, the participants also stated that, no, on its own, this permit does not. Participants generally felt that the CGP is too specialized. With a specific focus on construction-related pollutants and water quality, the CGP leaves other environmental issues unaddressed and does nothing to communicate the need for a comprehensive approach environmental and ecological protection. Issues related to endangered species and habitat protection, for example, are not addressed by the CGP, despite the impacts of construction and development to these resources. Additionally, participants felt that the CGP may only have the potential to communicate effectively with those working at higher project or programmatic levels or those who have already been somewhat educated in matters of the environment and sustainability. From this perspective, people living outside of the environmental bubble (to include the laborers and crew members actually performing the construction activities and thus directly responsible for controlling soil disturbance) may not see or perceive the same messages in their own exposure to the CGP. In this case, the rhetorical influence of the CGP is actually limited—limited to the people who make high-level decisions about schedules and budgets, while the people who actually come into contact with the environment by digging trenches and cutting asphalt remain unreached.

While this research shed light on the questions I asked, as is true with most research questions, more work is necessary before these questions can be more fully answered. The overarching question of "Does the CGP communicate the need for sustainability?" is complicated. The answer is that to some extent, at least, it seems to. Certainly on paper, the CGP makes use of rhetorical tools, relying strongly on the visual presentation of scientific data and a connection with over 30 years of federal water quality laws to justify its constraints. In practice, it appears that the CGP also promotes sustainable behaviors. In general, sustainable choices are made while destructive behaviors are eliminated in light of exposure to the CGP. Even when taking criticisms into account, it can be concluded that (both on paper and in real life), environmental regulations have the potential to communicate the importance of sustainability. And if these regulations are used and taught properly, sustainability can be transformed from a mere concept into things with the ability to be more tangibly measured: actions, behaviors, and choices.

Future research in this area should consider other types of environmental regulations as well as an expanded research group for case studies. Future studies should incorporate individuals working at all levels within a project or program, specifically people who don't have a long history of education and working with environmental regulations. This will enable researchers to see how the mind-set of compliance impacts those with less of a predisposition to regs.

References

1. A.M. Rugman, "Corporate strategies and environmental regulations—an organizing framework," in *International Business: Critical Perspectives on Business and Management*. London: Routledge, 2002, pp. 301–324.
2. M. Altman, "When green isn't mean: Economic theory and the heuristics of the impact of environmental regulations on competitiveness and opportunity costs," *Ecological Economics*, vol. 36, pp. 31–44, 2001.
3. M.L. Cropper, W.E. Oates, "Environmental economics: A survey," *Journal of Economic Literature*, vol. 30, no. 2, pp. 675–740, 1992.
4. R.A. Clarke et al., "The challenge of going green," *Harvard Business Review*, vol. 72, no.4, pp. 37–48, 1994.
5. "Environmental Protection Agency," *National Pollutant Discharge Elimination System (NPDES) Construction Site Stormwater Runoff Control*, January 12, 2007. Available: http://cfpub.epa.gov/npdes/stormwater/menuofbmps/index.cfm?action=min_measure&min_measure_id=4. Accessed on June 26, 2015.
6. "State of California Water Resources Control Board," *State Water Resources Control Board: Construction General Storm Water Permit*, July 1, 2010. Available: http://www.waterboards.ca.gov/water_issues/programs/stormwater/constpermits.shtml. Accessed on June 26, 2015.
7. State of California Water Resources Control Board (2009). Construction General Permit Fact Sheet. Available: http://www.waterboards.ca.gov/water_issues/programs/stormwater/docs/constpermits/wqo_2009_0009_factsheet.pdf. Accessed on June 26, 2015.

6

Influences of Technical Documentation and Its Translation on Efficiency and Customer Satisfaction

Elena Sperandio

6.1 Considering Technical Documentation

We all have experienced something similar. You bought a new television and you are not able to connect it because you do not understand the descriptions in the manual. You discover, as many people have, that in your multilingual manual, a procedural description in one language often differs from the procedural description in another language. Modern manuals typically accompany products that are distributed to multiple countries. Therefore, they must be given in multiple languages. Because those manuals include the same instructions in multiple languages, they are usually written in the primary language of the manufacturing company and then translated into the remaining languages. This can cause problems with procedures and terminology. In the worst case scenarios, descriptions become nonsense after translation.

Too often, users of technical documentation are confronted with these kinds of problems routinely and they face more serious matters than a new television. Unfortunately, the same translational problems that affect television manuals also affect food and drug safety information and descriptions. In this chapter, I will explain from my experience why this is so, what tools are available to help, and the resulting challenges faced by technical communicators.

Communication Practices in Engineering, Manufacturing, and Research for Food and Water Safety, First Edition.
Edited by David Wright.
© 2015 The Institute of Electrical and Electronics Engineers, Inc. Published 2015 by John Wiley & Sons, Inc.

A wide range of programs for recording and monitoring business processes—as well as for publishing, product data management, technical documentation, and translation—are currently on the market and are in use at many companies. No extensive publicly available studies, to my knowledge, have been conducted on the degree of diffusion of these programs on whether using them makes sense or on whether the companies concerned were able to improve their processes as a result. Therefore, there are no clear answers to these questions.

There is one aspect of monitoring and managing business processes that has probably been investigated more than any other: it is the introduction of enterprise resource planning (ERP) systems. ERP systems integrate internal and external management information across an entire organization, embracing finance/accounting, manufacturing, sales and service, and customer relationship management. ERP systems automate this integration with an integrated software application. Their purpose is to facilitate the flow of information between business functions inside the boundaries of the organization and manage the connections to outside stakeholders [1].

Therefore, in this chapter, all assumptions and the described consequences are based solely on my own empirical experience. My experience is based on 15 years of practice in the field of technical communication. Working for a translation agency that specialized in technical translations, I have gained insight into the content and in the technical systems employed for managing and generating content in the technical communication of various companies. My data have been gathered in cooperation with companies from various industries and industries of various sizes and largely comes from the field of technical documentation. The companies are active in various fields, from the food and pharmaceutical industries to agriculture and machine engineering. Their sizes range from 100 to 50,000 employees worldwide.

6.1.1 The Problem with Integrating Systems

Based on my experience, the degree of integration of the various systems (the availability of interfaces between the implemented programs) is very low. In the best case scenario, this situation only produces redundant data and increases costs. In the worst case, safety-relevant errors can occur and it can be difficult to find their origin. This chapter describes my experiences and my assessment of the situation in the hope of providing food for thought.

Depending on the department and the requirements, many different kinds of software are used in large companies. My interest here is to look at programs that were of strategic importance in those companies—those programs that either reflect business processes or that are used to manage data and content that are crucial to product quality and product safety, and therefore to the company.

Except for parts of ERP systems (e.g., accounting) almost all of these programs (PIM systems, CMS, and DMS) are used in the field of technical communication. Technical details, product specifications, and customer requirements are entered into and stored in these programs. Thus, they represent an important strategic tool for companies and the maintenance and integration of such systems is very important for product safety and efficiency.

The following systems are described briefly, with the focus on content management systems and translation management systems:

- ERP (enterprise resource planning systems)
- PIM (production information management systems)
- DMS/CMS (document management systems/content management systems)
- TMS/CAT (translation memory systems/computer-aided translation systems)

Next is an overview of the data management of technical content and the management of the corresponding translations. A description of the workflows in small, medium-sized, and large companies is given and the advantages and disadvantages of the respective solutions are discussed.

6.1.2 Enterprise Resource Planning Systems

In recent years, companies have increasingly started to use software that has been developed for simplifying work procedures within a company. Among this software, ERP systems are the most widely used and are aimed at ensuring needs-based distribution of resources across the entire company. These systems reflect business processes. A process sequence is defined company-wide across several departments and is then implemented within the ERP system. Implementation of such a system has far-reaching effects.

> For instance, Nestlé SA, which has its headquarters in Switzerland, announced in June 2000 that it intended to invest US$280 million in the introduction of a company-wide ERP system. Its 200 subsidiaries and branches in 80 companies were integrated into the system by 2004 and the system has been judged a success since that time [2].

It is estimated that 74% of all manufacturing companies use an ERP system [3]. Because ERP systems largely save and evaluate data that are relevant for management and control, these systems are more widely used than other systems, and, as the Nestlé example shows, they are more likely to gain financial support from upper management than more specialized programs.

Historically, the development of ERP systems started with accounting systems, which were later supplemented with inventory management and supplier systems. ERP systems are usually database-driven. The information is saved in tables, which can then be accessed company-wide using a client-server architecture. As the amount of processes (and thus data) that is being managed is constantly increasing, other software solutions have been introduced in parallel to the company-wide ERP systems to support other processes and tasks within the company.

6.1.3　Production Information Management Systems

Usually, ERP systems have not been designed to include all product data such as product descriptions, images, and technical data for the product or parts of it (such as ingredients). To meet this need, PIM systems have been developed in parallel. Within these systems, product data are recorded and can subsequently be published in catalogs, webshops, and websites.

The degree of diffusion of PIM systems in the mechanical engineering industry can currently only be estimated. However, it can be assumed that internationally active companies with several production sites, a large product range, and electronic sales channels cannot refrain from using such systems.

PIM systems are usually employed across several departments, subsidiaries, or branches, (e.g., engineering, marketing, and sales departments). PIM systems, like ERP systems, are usually database-driven; the product data (technical descriptions, dimensions, etc.) are stored in the database. However, the difference between PIM and ERP systems is that in PIM systems the stored data serve not only internal purposes but also production processes. The data are also used in published media such as catalogs and websites. Therefore, in contrast to ERP systems, PIM systems are designed to hold multilingual content, as the catalogs and websites have to be published in respective national languages. The challenges that this can cause will be discussed later.

6.1.4　Document Management Systems/Content Management Systems

Legal regulations and the increasing complexity of industrial machines and production processes have significantly increased documentation demands. The documentation volume has increased and has led to the need to manage the existing documentation and to achieve a high level of reusability; that is, verbiage and even layout need to be used and reused in order to save money for future releases of any product or process.

The two terms DMS and CMS are today used interchangeably. There is no clear distinction in their market use. The abbreviation DMS is also frequently interpreted to mean data management system. In this chapter, DMS is defined as a system that manages files, whereas CMS only refers to management of content.

DMS systems for technical documentation manage the files directly in the format in which they will be published. The files are indexed to enable searching by keywords or version numbers. However, adding metadata to the files is only possible with limited scope and is very complicated. The size of the individual files is determined by the documentation requirements. Images and other language versions are managed in the same way. If a file must be changed, it is "checked out" and then locked for editing by other persons. This ensures that the data sets remain synchronized. A large disadvantage of a DMS is that the file format is usually proprietary and thus can only be manipulated within limits. New versions of the implemented software are sometimes no longer compatible with the DMS, which leads to constant adaptations.

CMS, on the other hand, has the advantage that it usually implements an open standard (XML 1.0 or XML 1.1). The need to use metadata for identification has been foreseen in the format itself and the information is available in a structured form. This ensures that individual parts of the technical documentation can be easily found and reused. A fundamental characteristic of all markup languages is that the content and layout are separated, so that the authors need not concern themselves with the layout of the documentation when writing it.

Despite the advantages of this separation, the degree of diffusion for CMS/DMS systems can, as with PIM systems, only be estimated at this time. However, large internationally active companies can surely not cope today without using appropriate supporting software. In my experience, companies with less than 1000 employees usually do not use a system of this kind, whereas a company with more than 10,000 employees cannot afford to do without a CMS or DMS. The use of DMS or CMS is usually limited to a single department at a single location.

6.1.5 Translation Memory Systems/Computer-Aided Translation

Parallel to DMS/CMS, computer-aided translation tools first became available at the end of the 1990s. In the beginning, those tools were usually called CAT tools or translation workbenches. As of this writing, the term "TMS" is more common.

The heart of a TMS or CAT tool is the translation memory. This memory contains the respective source text and its translation. To ensure precise matching of the original text to its translation, the source text is segmented, usually into linguistically sensible units such as sentences. These segments are then translated and the pair (consisting of the source sentence and its translation) is saved.

A translation memory is always bilingual. When a new translation is needed, the translation memory is searched for identical or similar text that has been translated in the past. If there is a match, the previous translation is suggested to the translator. For example, let's imagine a manual that states, "The ball is red." Upon translation for the second version of the manual, the sentence was translated as "The ball is blue." The translator pulls the prior sentence construction and information from the system (that a similar sentence has been already translated), and it uses that match in the translation memory. The system will also inform the translator about the match rate—which is the percentage of similarity.

A TMS does not provide fully automated translations. The translations are always checked by a human translator. In the beginning, this software was used only by the translators themselves, but the technology has evolved and today it is also increasingly being used in companies. The motivation for doing so differs greatly. The first companies that invested in a TMS had their own translation departments for technical documentation and thus profited from the new technology. Today, data exchange with service providers, management of translation memories, and processing of translation projects are also becoming focus points. Therefore, the term "TMS" has become more common in recent years.

In my experience, only very large companies implement TMS; this is in part due to the relatively high purchasing costs.

In principle, computer-aided translation and translation memory systems are the same technology. However, even though the number of companies that are using a TMS is increasing, it is still not a widely used technology.

6.2 Data Management in Technical Communication

It is not possible to accurately describe the "normal" process of how data are managed in the engineering industry, as the use and integration of various solutions differ significantly depending upon the company. Similarities can best be identified when the companies are categorized by their size, because companies of similar size have similar software requirements.

One reason for this is that relevant programs such as ERP, PIM, and CMS are first and foremost designed for referencing existing processes within companies, with actual content being referenced to a much lesser degree. ERP and PIM software are usually designed to have different types of data interlinked (e.g., offer data for a time schedule of a project). Plausibility checks (e.g., if the start date of a task cannot be after the end date) are all inserted as order numbers but are not implemented. It depends solely on the user to ensure that the entered data are correct. Even in CMS software, which manages content, the focus is on how the content is managed and the level of reusability, not on the content itself. So just interlinking the data of several deployed systems is not sufficient. The quality of the data still needs to be checked. Currently, there are tools for author assistance and terminology tools in use which help, to a certain extent, ensure quality (e.g., terminology consistency). These tools are used mainly in the technical documentation department but not in the engineering or manufacturing departments of the company. This situation causes confusion and shows how important and safety-relevant a sensible integration of all data can be.

6.2.1 Development and Diffusion of Data Management Tools

With computers today being commonplace in most fields of work, specialized software applications for all imaginable areas are now in demand. The majority of the systems discussed here were developed in parallel. Therefore, in most companies, the degree of integration of the systems is relatively low. The oldest among these are the ERP systems. They evolved from MRP solutions (material requirements planning) from the 1960s. Since the 1990s, they have become increasingly common. Today, they are designed to integrate internal and external management information across an entire organization. Later, as new technical possibilities became available, PIM systems as well as DMS and CMS solutions were developed, all of which are constantly gaining importance in industrial use.

PIM systems refer to processes focused on centrally managing information about products. They have become a necessity for handing over product data from one department to the next, starting with the design data. This is vital for modern manufacturing processes and warehousing systems (e.g., "just-in-time" production), which are hardly possible anymore without such systems. Furthermore, large companies are increasingly placing their orders from suppliers electronically, which requires catalogs, item numbers, and product specifications to be available online. PIM systems are able to load the product data directly into other electronic publishing solutions (e.g., electronic catalogs, webshops, websites), which bears the risk that product data are published without having been verified beforehand. Consequently, the risk of publishing inconsistent or even wrong data arises.

As the processes covered by the PIM and ERP systems are interlinked, it can be assumed that the degree of integration of these two systems is high. At roughly the same time as the introduction of PIM and ERP, a solution for another problem was needed. Due to changes in [4] legislation (especially in Europe, with the introduction of Directive 98/37/EC of June 22, 1998, revised as 2006/42/EC of June 9, 2006), manufacturers of machines, including machines used to manufacture food and pharmaceuticals, are obliged to produce an operating manual for operation of the machine. In accordance with the directive, this operating manual must be translated into the language or languages of the country where the machine will be used. Due to these requirements and the increasing complexity of the machines, the volume of machine documentation has increased significantly.

In the 1990s, technical documentation was mainly produced in DTP (desktop publishing), because these programs were the only tools that allowed text and images to be integrated and published with an acceptable quality. Due to the ever-growing documentation volume and the fact that file-based programs are used, file management became necessary. However, the first systems that were implemented had not been developed especially for technical documentation. Initially these systems were intended for data archiving. Therefore, the focus was on locating the files again, not on reusing them or implementing version control. Small software companies recognized the need for more specialized solutions and developed these systems to match the requirements of the technical documentation departments.

The introduction of XML in 1998 caused a paradigm shift in the field of technical documentation, making it possible to manage content that need not necessarily be stored in a database. However, although this new standard made content management easy, publishing the data became a problem, as a transcription into a printable format was needed. Additional applications for publishing XML data had to be developed. However, this meant costs and effort that only large companies were able and willing to invest. Solutions that utilize standard applications for publishing have only come into existence in recent years. The only disadvantage of using standard programs such as Microsoft Word® or FrameMaker® is the fact that the company cannot be sure that the content saved in the CMS is still "single source" (that the data in the CMS and the published documents are absolutely identical) as the published manuals are editable.

At roughly the same time as the first PIM, DMS, and CMS solutions, the first TMS was developed—not as a result of the needs of the industry, but as a niche program for the field of technical translations. The programs that are today called translations memory systems were developed in the mid-1990s. The reason for this development was the high amount of redundant text in technical communication. Publications in technical communication are characterized by many repetitions and a high level of reusability. This especially applies to the safety and maintenance chapters of an operating manual, but also for catalog texts. Furthermore, these texts have a simple syntax, as they are largely aimed at conveying information.

This fact is utilized by CAT programs that divide the text into segments and save the original segment with its translation as a pair. During a new translation, fuzzy logic algorithms scan these saved pairs and try to find matches or similarities between the previously translated text and the new text. If a match is found, the existing translation is suggested to the translator. Thanks to higher performance and in particular new formats, new possibilities have become available. Today translation memory systems not only manage translation memories but also enable quality checks, terminology management, and automatic replacements. The newest developments are in the area of translation project management and data exchange.

From my experience, manufacturing industries and the food industry are among the more traditional and conservative and tend to only implement innovations once these have found a permanent place in other fields. Analogously, it is usually the larger companies that risk investing in a new system, whereas small and medium-sized enterprises tend to wait.

Today, ERP systems are widely accepted and even PIM systems enjoy relatively widespread acceptance across all industries. This is largely because, depending on the system, different people in the companies are responsible for making the purchasing decisions. However, this also influences the available budget and the acceptance for introducing changes.

For ERP systems, the top management is usually responsible. As a result, the budget is usually only limited by the means of the company. Those responsible for PIM systems are usually also in high strategic positions in the company, such as engineering, marketing, and sales, and thus have a high influence on investments. Therefore, the budget for and the willingness to use a new system are often high.

The technical documentation department, on the other hand, usually only occupies a small niche in the company. Accordingly, the budget and the willingness to invest in innovations in this area are low. From the statements above, the following can be derived:

- ERP and PIM systems are the most widely used.
- DMS, CMS, and TMS solutions are generally only used by large companies.

The next section focuses on the technical documentation department and thus on DMS and CMS. In addition, TMS and problems related to translations are reviewed. To get a better overview, the description follows the previously mentioned categorization based on the size of the company:

- Small company (>500 employees)
- Medium-sized company (>10,000 employees)
- Large company (<10,000 employees).

As mentioned above, experience shows that the industrial sector does not play a decisive role.

6.3   Technical Communication in Small Companies

Aside from a few exceptions, small companies do not have a professional technical documentation department. That lack of professionalism applies, on the one hand, to the training and qualifications of the employees and, on the other hand, to the use of tools.

Over my years of work, I found that employees usually come to the documentation department from other departments and have no training or experience in professional writing or in working with relevant software. Therefore, these companies usually have no style guides for their manuals. If any terminology management is done, it usually is done with a basic list. For creating the documents, standard Microsoft Office® applications that are already available in the company are used. The willingness to invest financial resources into this department of the company is very low. Technical communication is managed by means of the file system of the operating system that the company uses. The responsibility to design the structure usually lies with the head editor of the technical documentation department.

A typical department in a small company can look as follows: The documentation team consists of two persons, of which one only works part-time. Manuals are written in Microsoft Word®, the spare parts lists are compiled in Excel®. Images are created with Adobe Illustrator®. No quality assurance is in place. The files are saved in the normal file system, in folders labeled with the machine designation. Translations are saved in subfolders with a language ID as defined in the ISO 639-1 standard [5]. All lists are in a separate folder. There is a bilingual terminology list (e.g., German-English), which is exchanged with other departments at regular intervals, but not synchronized. The sole responsibility for the structure, layout, and content of the technical documentation lies with the head editor. No regular updating is done.

Also, there is often no advanced planning for document creation. Frequently, the technical writing department is not informed if a machine is ordered. The request for the required machine documentation is usually only communicated after the machine has been completed and is ready to be shipped. As no content management is conducted for the existing documentation, the technical editor must judge which of the existing manuals might be suitable as basis for the new documentation; the same applies to the graphics. The editor must acquire the technical information regarding the new machine "on site" in the production facility.

The quality and on-time delivery of the documentation depends on the work of two persons in the company. This fact can have a great influence on product safety. It is sufficient that one person makes a mistake and the consequences can even be life threatening for the product consumer. Just imagine a wrong description inside a booklet on a pharmaceutical product!

6.3.1   Workflow Advantages in Small Companies

The advantage of this organizational structure and workflow lies largely in the very low initial cost. No new software and hardware need to be purchased and usually no training costs are incurred. Also, the IT department does not have to set up a new infrastructure.

As the technical editor usually worked in the company before being assigned to the documentation department, that person is an expert and knows the product and all product versions very well. Therefore, the technical documentation department can react quickly and flexibly to machine changes and adapt the documentation accordingly.

6.3.2 Workflow Disadvantages in Small Companies

The obvious disadvantage of small-company workflow is that the entire technical communication of the company depends on two persons. Frequently, these persons are not qualified technical writers but have been reassigned from other positions at the same company. Therefore, their knowledge regarding legislation, regulations, and standards depends on their experience and their own initiative. Knowledge regarding the use of the respective software tools has usually been acquired through learning-by-doing. Frequently, important functions of the used software are unknown or not adequately understood. Furthermore, there are no style guides or similar instructions for creating the documents.

Existing documentation can only be found and reused to a limited extend. Frequently, due to the use of Microsoft Office^{TM} programs, one manual is equivalent to one file. To save time and money, the editors often try to combine parts of older translated manuals to create a new manual that is already in the target foreign language. However, "pasting together" old manuals is very error prone. It is also not possible to objectively check the extent to which an old manual is suitable for reuse. The choice is based on the memories of the editor.

As no defined processes and no tools are available for creating the documentation, the efficiency and quality of the documentation cannot be measured. It is also not possible to integrate the technical documentation into the other systems that are used in the company, as appropriate software is not available. The technical communication with the other departments is, therefore, conducted informally and often depends on the sympathies and antipathies among the employees.

6.4 Technical Communication in Medium-Sized Companies

Medium-sized companies increasingly tend to standardize their processes and, therefore, to invest in the corresponding programs and systems to make the work processes more verifiable and efficient. In recent years this development has also spread to the field of technical documentation and has led to large changes. The technical documentation personnel are usually qualified technical editors and writers.

The technical documentation department usually consists of 5 to 10 employees who have specific duties. There are the authors that write the documentation, and the project managers who are responsible for managing document creation, publishing, and translations. Normally, these departments utilize a large number of different tools, from programs that aid the author to TMS and DTP programs. content management systems have typically been in use since it became possible to publish by means of standard applications. Usually, technical communicators attempt to find a content management solution that is based on or functions as an add-on to an existing DTP program to keep the costs low.

The documentation department of a medium-sized company might include four authors who are involved with creating the documentation, compiling spare parts lists, and creating images and graphics for the manuals. They use finished templates and previously written blocks of standard text from the CMS. For instance, a standard text block might contain general safety instructions. The documentation is written with an XML editor included in the scope of supply of the CMS. Additionally, programs that provide support to the authors are used. These programs ensure that style guides and the specified terminology are observed.

Two project managers work on planning and publishing the documents that have been or are to be created and order the required translations. The project managers decide how to monitor deadline adherence and manage the orders. There is no standard solution for project planning in this field. The integrated project functions in CMS. Lists and the integrated project planning module of the TMS are all used in parallel. The TMS is used to manage the terminology and translation memories for all languages. The project manager uses this data to check the offers of external translation providers.

The documents are published in FrameMaker^{TM}. With this method, a printed version and a PDF version are generated. The PDF can then be made available for download or burned onto a CD. Most processes have been described and are verifiable. Terminology and style guides are available, as well as programs that check for correct usage. A quality system has been implemented. Thanks to the high level of support and the defined processes, colleagues can be trained quickly to take over tasks.

6.4.1 Workflow Advantages in Medium-Sized Companies

A large advantage of this work method is that almost all processes have been described and can be monitored/tracked. The processes can thus be supported by appropriate software. In addition, errors are traceable, which is very important for the quality of the documentation and the safety during use of the respective machines.

By using appropriate programs and training the technical documentation employees, a high degree of efficiency is achieved. The use of modern CMS allows a high degree of reuse of previously written text. This reduces the work involved in creating a new document and increases the quality of the documentation, as text that has already been validated and proofread can be reused multiple times. Additionally, programs that provide author support are used. These ensure that the specified terminology is used and the style guides are observed. This safeguards uniform style and quality throughout, even when the documentation is being written by several authors.

In addition to the CMS, a translation memory solution is used. It largely handles the same tasks for the translations as the CMS handles for the authors. The translation memories are exchanged with the translation provider(s). Content thus becomes easier to compare and check, which results in better cost calculation and control. The translation memory exchange allows the company to quickly and easily switch to a different service provider without incurring large losses, or to divide large projects among several service providers, if necessary. In addition, TM solutions can lower translation pricing because of the following:

- Cost and time schedules for translations are usually established by analyzing the text to be translated against the translation memory in order to verify how many matches and/or repetitions the text contains.
- Those matches usually cost less than a totally new translation—especially when there is a high match rate.

6.4.2 Workflow Disadvantages in Medium-Sized Companies

The largest advantage of this work method is simultaneously its largest disadvantage. First, the purchasing costs are high. Second, the cost and effort required for the IT department to maintain the software can be significant, because these programs are usually client-server applications that cannot simply be installed on a computer. Additionally, to ensure that all employees of the department are capable of using the software efficiently, training must be conducted. Apart from the client-server systems, a desktop-based software is used for publishing. For desktop publishing programs there is usually no network solution. Therefore, they must be installed on individual workstations. Yet, the largest problem in this case is the editability of the content. At any given time there is no guarantee that content stored in the CMS matches content that has been sent to customers.

Redundant data are also a problem, because few or no interfaces exist between the applications. The data sets that are stored in the different systems (CMS, author support, TMS) partly overlap. As these systems are not integrated, no data synchronization takes place. As a result, data sets that were initially identical start to diverge over time. Cleaning up these differences at certain intervals is complicated and always results in data loss, as the diverging data in the lower-priority system is simply deleted.

In addition to the problems caused when the software used is not integrated even within the technical documentation department, there is also the matter of the lack of integration with other systems used in the company, such as the ERP or PIM, which are also often part of the technical communication of the company.

In spite of a high degree of automation and high strategic importance, there is no standard for project management. Punctual delivery of high quality documentation has a large effect, as the documentation represents a direct interface to the customer. High quality documentation that is delivered in time to the market means a high safety standard for the product because the content of the documentation is checked by a quality assurance system and validated contents are stored and reusable. It is crucial to product safety to have the technical documentation or product description together with the product delivered.

6.5 Technical Communication in Large Companies

In large companies, a differentiation must be made between two strategies for organizing technical documentation. Some companies choose full decentralization of their documentation. In other words, each production facility has its own documentation department. Depending on the size of the production facility, the documentation department corresponds to that of a small or medium-sized enterprise, with the same advantages and disadvantages. A disadvantage of this work method is that within the same company the presentation of documentation can differ greatly from factory to factory and from country to country. This can be a problem for the image and corporate identity of the company.

The second strategy is to centralize the documentation at a few locations. This makes it possible to define and standardize processes, to ensure similar documentation and thus a similar company presentation worldwide. The technical documentation staff consist largely of technical writers, editors, project managers, and translators. Software is usually standardized for the needs of the company. The degree of customization of the implemented solutions is, therefore, frequently very high. However, the department is usually spread across several locations, which puts additional demands on the software. The software must be Internet-capable and must have appropriate security features to prevent unauthorized access to the intranet of the company. The authors usually work at different locations, usually at the factories in which the machines to be documented are being built. The project managers and the company's internal translators usually work together at a single location. CMS, TMS, author support programs, image editing programs, DTP software, Microsoft Office applications, and project management software are used. But large companies are forced to manage large amounts of data, which can bring standard solutions to their limits. For example, it is not unusual for a single publication to have 1.5 million words. Therefore, a typical workflow often looks like this:

1. The authors document new machines developed by the factory and the updates made to existing machines.
2. To this end, the style guides and templates saved in the CMS, as well as previous versions of the document and so-called standard blocks (blocks of text that contain regulations, safety instructions, and other information that is not machine specific) are used.
3. The CMS saves the data, with the associated keywords, in XML format. Additional information such as a unique ID, date, author/editor, and processing status is also stored. The size of the individual blocks depends on the topic area (see Figure 6.1).

Even if the company has a corporate language (language that is used for internal communication worldwide in the company), not all authors write the documentation in this language, as they do not have adequate knowledge of the corporate language and the quality of the documentation would be negatively affected. To enable better management of the entire documentation, these blocks are then translated into the corporate language. This first translation is usually performed by an in-house translator of the company and is not given to an external provider.

The authors use XML editors to write the documentation. The diffusion of additional tools, such as author support tools and terminology solutions, differs from site to site,

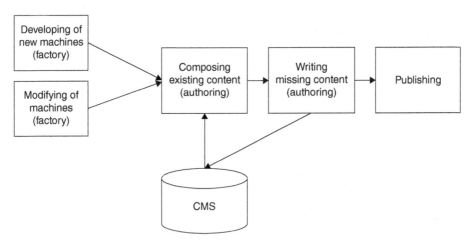

FIGURE 6.1. Typical technical communication workflow in a large company. These processes are often affected by word counts that push systems to their limits and differing physical locations for writers.

as these programs are not integrated into the company-wide CMS and TMS. After the documentation has been created, it is published by means of a publishing system. After the documentation has been completed, the factories request the required translations at the project management of the documentation department. The project management department compiles the data for the translation, looking for the blocks that still need to be translated in the CMS. If this is not the first time that a manual is translated into the language concerned, the compiled data will not include all components of the manual. The project manager then gives the translation job either to an internal translation department or to an external provider. The translation jobs are distributed by means of software developed especially for the company, which assigns the dates and deadlines for the job, sends the data, and monitors the job.

In addition to the manuals, which are drafted and edited in accordance with defined processes, there are document types that are not included in the CMS. This, for example, includes software control texts for machines and text that occurs in images. When creating these documents, the authors use the standard solutions that are available in the company. These are selected based on the knowledge, preferences, or "traditions" of the respective department.

6.5.1 Workflow Advantages in Large Companies

As described in the section on medium-sized companies, larger companies have the advantage of using defined processes and thus quality management systems. The same criteria can be applied across several sites worldwide, providing a uniform standard across borders. This also makes it easier to make use of external authors. A uniform standard is achieved by using large TMS for the translations.

Using many solutions developed for the company makes efficient and structured work possible even in areas for which there is no standard software on the market. This includes publishing, for example. The large advantage of using a publishing system, instead of a DTP system, is that the resulting document edition from the publishing system is usually no longer editable. This guarantees that the supplied manual matches the one saved in the CMS. The project management also has professional software for monitoring the projects at its disposal. It is, therefore, easy to monitor deadlines and hand over projects from one manager to another.

6.5.2 Workflow Disadvantages in Large Companies

The costs for purchasing and maintaining the systems, training the employees, and backing up the data are high. In addition to the actual systems, database servers have to be installed, the individual workstations have to be secured against malware, viruses, and so on, and high network capacities are required to process and exchange the data volume within an acceptable time. Furthermore, developments that are in part proprietary are used. This means that data exchange is not always easily possible. This can lead to huge dependencies, as access to a company's internal data are only possible by means of a specific software. This clearly shows that there is a high dependency on the IT systems working flawlessly. If these systems fail, no work is possible. Therefore, in addition to a backup, a certain level of system redundancy must be ensured.

In addition to the technical requirements, employees must work with a high degree of professionalism and have knowledge of many different programs. Using solutions that have been especially developed or adapted for the company requires intense training, because in most cases, no extensive documentation or help files exist. This means that both IT and personnel costs are high.

As is the case in medium-sized companies, the use of many different systems presents a big problem. These systems manage partly identical data sets, yet cannot easily be synchronized. The use of single workstation software that is not integrated into a system further complicates matters. And last but not least, the systems used in the documentation department are not usually integrated into company-wide systems. This is more critical in large companies than in medium-sized companies, as a larger number of persons are involved with the data creation and editing and thus a larger divergence between the data sets within a short time is possible.

6.6 Translation of Technical Information

As is the case with CMS, little or no research has been conducted on the implementation and organization of translations or on the software used for managing and creating the translations.

Software applications for managing and producing technical translations are niche products, therefore only rough estimates can be made regarding their diffusion and use.

However, over the course of our many years of experience at our translation company, we came to realize that certain similarities can be identified between companies that have roughly the same size. This, at least, applies to all companies that require extensive technical documentation and have a high translation volume. Therefore this section, like the previous, will also be structured by company size.

6.6.1 Translations in Small Companies

Generally speaking, as the requirements and the documentation volume increase, the translation volume also grows. Over the course of years, more and more target languages are added. For many of our clients, the technical translation department in small companies grew historically. Frequently, the translations are still performed by individual freelancers, with the result that a different translator must be found for almost every target language. For this reason, many companies started to make use of translation agencies or offices, as these can usually provide all required target languages.

Technical tools (e.g., project management software, translation memory systems) are rarely used by small companies. This applies both to the management of translation jobs and to the maintenance of terminology and translation memories. Some editors use and update Excel lists for the most frequently used target languages. However, these are not synchronized with any other system within the company. It is up to the translators or translation offices to implement appropriate technological solutions for the benefit of the company.

It is rare for quality control of translations to exist in a small company. Again, the responsibility usually lies with the supplier. Frequently, due to time and cost constraints, editors attempt to paste together an updated document from old translated manuals without being competent in the foreign language concerned. The translator then only gets sentences or sentence fragments that have been taken out of their context. The resulting translations are subsequently inserted without checking whether the terminology is consistent with the rest of the manual, or whether the translation is appropriate for the context. Figure 6.2 shows a standard manual translation sheet.

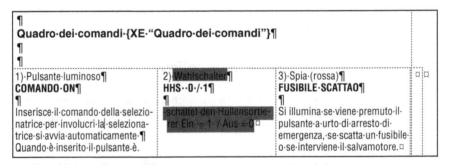

FIGURE 6.2. A standard translation sheet used by small companies. Such tools offer basic translation but often do not incorporate contextual elements or typesetting features that make the difference in true translation.

Another challenge is foreign character sets. The issue of displaying foreign characters was solved when Unicode (standard regarding character encoding) was introduced, but this does not solve the problem of foreign languages having other typesetting rules. This concerns syllabication, citations, bold emphasis, line breaks, and much more. Being proficient in the use of the software, therefore, does not also mean being able to generate a professionally typeset document in the foreign language.

Translation Workflow Advantages in Small Companies   One of the most important advantages of working with individual translators or small translation offices for a long period is that the translators become very familiar with the topic and the requirements. They know the machines and rarely have to ask questions. This accelerates the translation process and usually guarantees results that are of acceptable or even high quality.

The purchasing and documentation departments frequently claim that this working method is less costly. As no generally available data exist regarding the translation quality and automation through the use of modern software as compared to working with individual translators, and the workflow of the translators can differ significantly, this claim cannot be proven.

Translation Workflow Disadvantages in Small Companies   One disadvantage is without doubt the higher administration effort. If documentation must be translated into several languages, the required effort is multiplied. Additionally, there is the matter of quality problems. High quality in one language does not mean that the quality is equally high in the other languages. For example, if one translator asks a question about a text section and another does not, there is reason to doubt whether the translator that did not ask the question understood the section correctly.

If deadlines are tight or order volumes are high, another problem can be the availability of the translator. In that case, the company might have to use a completely different translation supplier. As a consequence, the translations will not be consistent and probably differ considerably in quality.

Yet another problem is making the data available. If, instead of standard Office programs or open standards (such as markup languages), DTP software is used, it cannot be assumed that the translator or a smaller translation office will have the software and will know how to use it. If not, translators frequently try to extract the text, in order to provide the translator with only the text itself. However, this harbors several dangers. First, the translator cannot always be certain whether the text fragment is a heading, a list point, or a description that in some cases would each require a different translation. Second, exporting text from these programs is usually fairly easy, but importing it back into the respective program presents a bigger problem. Frequently, the only option is to manually copy the text section by section. This can be very time consuming and error prone, especially if the person who does this task cannot understand the language concerned.

Even when the translator is able to work with the original format directly, the problems presented by foreign language typesetting are not solved. Most translators are not DTP specialists and do not know the respective typesetting rules. As a result, the technical editor in the company is responsible for preparing the data for printing. Depending on the format and the knowledge of the editor, this can be very costly and time consuming.

6.6.2 Translations in Medium-Sized Companies

Many medium-sized companies are characterized by their expertise and accordingly by high quality standards where their products and the image of the company are concerned. This is also reflected in their willingness to invest in areas such as the technical documentation. Most companies use translation memory systems, yet with different degrees of automation. Some only use workstation versions of the TMS software for managing the translation memories and for checking the offers of the suppliers. Others have set up large proprietary client-server systems.

In recent years, it has become increasingly important to introduce and manage terminology not only in the most common languages but across all languages. Compiling and maintaining terminology data is very complicated and time-consuming. Therefore, the scope, the distribution, and the quality of the terminology differ greatly from company to company.

Very few medium-sized companies still work with freelancers. Rarely can the requirements of the companies be met by individual translators. Usually the companies work with one or two medium-sized offices or agencies that can provide all required language combinations and can also offer all additional services beyond the translation itself. It is taken for granted that these suppliers use translation technology and have a quality management system in place.

Additionally, the companies attempt to introduce processes into the translation that safeguard and improve the quality. As most companies have foreign branch offices or sales offices in the target countries, it is common to conduct a proofreading at the branch/subsidiary. Approval of the documents is then subject to this proofreading.

Translation Workflow Advantages in Medium-Sized Companies Using a TMS in the company makes it possible to check the translation volume and the overlap with existing translations, instead of trusting the word of the translation office. Thus, the costs and time required can be estimated in advance. By means of these programs, it is possible to check the formal quality of a translation. In this context, formalisms are aspects such as completeness, correct punctuation, spelling, correct numbers, correct tags, and adherence to predefined terminology. Using the TMS, data sets can be exchanged with different service providers and can thus be synchronized at regular intervals. The quality control for the content is done during internal proofreading at the branches/subsidiaries, where the proofreader checks that the translated text is suitable for the target market.

Translation Workflow Disadvantages in Medium-Sized Companies Using a TMS and managing memories and terminology requires time and effort, as well as specially trained personnel. Furthermore, the IT requirements and the purchasing costs (depending on the selected solution) can be very high. Apart from the need to provide the appropriate resources, another problem is the relatively high data redundancy. The data sets in the different systems partly overlap and cannot easily be synchronized.

An additional challenge is that the persons who are doing the proofreading frequently do not understand the source language and do not have training in the field of translation/documentation. In most cases, due to cost reasons, the branch offices do not have a TMS to check terminology and compare older, approved translations. This means that the proofreader does not know the terminology or the translation memory and, as a result of this lack of knowledge, perhaps changes fixed terms or approved translations. This also raises the question of which format should be used for the corrections. In most cases, an editable PDF is used and is commented by the proofreader. In the best case scenario, this PDF is then sent back to the translation office where the changes are then entered into the translation memory. Here, any inconsistency with the memory can still be detected. In the worst case scenario, the project management team enter the changes into the document and thus bypass the memory. With such complications, the question of whether internal proofreading at the branches should be viewed as an advantage or rather as a disadvantage remains unanswered.

6.6.3 Translations in Large Companies

In companies that have largely centralized their documentation and thus also their translations, large TMS client-server solutions are a matter of course. Furthermore, the manufacturers of the systems provide custom adaptations to the special needs of the respective companies. These adaptations first and foremost concern automations with which a translation job is automatically generated every time changes are made in the CMS. In general, this process can be described as follows:

1. After documentation has been created or updated in the CMS, the required language versions are requested.
2. The CMS compiles blocks of text that still have to be translated for the respective language and checks them out of the system.
3. When the checkout is performed, translation projects for the requested languages are generated automatically.
4. In the course of generating the projects, a project memory is created. In other words, the entire translation memory is searched for entries that have similarities with the text to be translated.
5. This project memory is then exported.

The same is done with the terminology stored in the system. The data to be translated, together with the project memory and the terminology, is then combined into a package that is sent to the selected service provider. The planning data (volume, delivery date, other notes) is usually transmitted by means of a project control module integrated into the TMS. The project control module generates a control file, which is then sent with the translation package. Once the translation is received from the provider, the translation

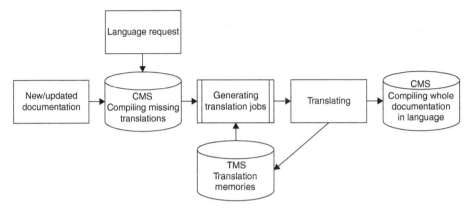

FIGURE 6.3. Translation workflows in large companies. While typically more thorough than in smaller companies, these content management systems are not yet capable of providing seamless translation. Quality assurance checks are still needed.

memory and the terminology are synchronized and the translated data are imported back into the CMS. Figure 6.3 represents translation job flow in large companies.

Depending on the solution used, the processes can differ slightly. For example, the request to generate a proof copy for the translator is not easily met, as the documentation must be reconverted from the translation memory format to the format that is used by the CMS and then must be sent through the publishing system. Therefore, quality assurance is not always provided for in the workflow. Furthermore, (formal) quality checks are defined within publishing systems. The project cannot be concluded before these checks have been performed.

Branch-office proofreading is usually not done, as the complex workflow and the short processing times make it almost impossible to integrate the proofreading and corrections.

Translation Workflow Advantages in Large Companies The high degree of automation ensures that all required information and data are reliably exchanged between the customer and the translation service provider. The reusability of translations is high and there is a terminology database that is always available. The automatic triggering of translation jobs and the automated compilation of translation packages eliminates most of the errors that can occur during the project assignment process.

Another advantage is that the data at the company are regularly updated each time a translation is delivered. However, the biggest advantage is that the TMS is at least partly integrated into the CMS. The CMS does not only manage the original manuals but also the translations. Thanks to the CMS organization, only those parts of a translation that have not yet been translated are sent to the translators, yet simultaneously a high degree of consistency in the translations is ensured.

Translation Workflow Disadvantages in Large Companies The purchasing costs for the standard solutions and the special adaptations, as well as the requirements to the IT department, are even higher than in the case of the medium-sized companies. Furthermore, these solutions are frequently so complex that their introductions can take years. Another aspect that must be considered is the fact that the software houses frequently promise solutions that cannot be implemented as is (e.g., performance, customizations, costs), which means that the companies frequently have to enter into compromises in order to not risk the entire project.

In spite of all the cost and effort invested and the complexity of the workflows, these workflows are also not integrated into the general company software. The foreign language data contained in the documentation are not used to create catalogs or marketing material. Likewise, the terminology used in the PIM by the marketing and engineering departments is not used in the documentation.

6.7   Consequences for Technical Communication

In spite of the use of new technology and the attempts to describe and standardize workflows, complete integration of the processes of the technical communication department within a company is not possible yet. The core problem is that technical communication and production processes are not integrated. Even when large companies are willing to invest large amounts of resources, they do not often integrate technical communication into their core processes. This poses the question of what effects these facts have on efficiency, customer satisfaction, and product safety.

Reliable data concerning the degree by which the use of a CMS in combination with a TMS increases the efficiency of the technical communication department does not exist. This, of course, also depends on the type of the documentation and numbers of revisions (frequent creation of similar documents can make better use of the reusability provided by a CMS and TMS than is the case with documents that have only few similarities). Therefore, claims regarding the percentage by which the use of certain systems can reduce the costs are not viable if the type of documentation has not been examined beforehand. Another aspect that determines whether using modern systems will lead to a significant increase in efficiency is the service life of a document. The longer individual documents are used, the more they profit from potential reusability.

Measuring customer satisfaction is a problem for all companies. Expectations differ depending on the industrial sector, document use, and price structures. Accordingly, the importance assigned to the technical documentation varies. The more complex the products, the more important their descriptions become. Who has not experienced the frustration caused by the bad operating manual of a DVD player or similar device? The same applies to operating and maintenance manuals. If these manuals are badly structured and the use of terminology is inconsistent, the results can go beyond mere customer dissatisfaction. The consequential operating errors can even be dangerous, which could be fatal for the manufacturing company. Safety measures may be scattered all over the manual and hard to find for the operator. Worse yet, large equipment documentation may fail because wording for an emergency "off" and emergency "shutdown" are used synonymously without explanation. Good translation ensures a safer work environment.

6.8   Assumptions About Technical Communication

As mentioned earlier, no reliable data are publically available. Therefore, the following conclusions are assumptions based on my many years of experience in the field of technical communication and translation. They are intended as inspiration for investigating this field and its position within the company more extensively, in order to achieve improvements for the company, the service provider, and the customer.

In the business world, "efficient" means that the costs for a product to be manufactured do not exceed the profit that can be achieved. The lower the costs and the higher the profits, the larger the efficiency. However, because production processes and workflows are becoming increasingly complex, it is also getting increasingly difficult to assess whether a workflow is efficient. This becomes even more difficult when processes that do not support the production itself are concerned (such as technical communication).

In this context, it is probably easier to determine what is *not* efficient. A core aspect that affects the efficiency of technical communication is the redundancy with which the data are recorded and managed. Each department in the company using its own solutions and programs that are not networked with each other is inefficient.

Such considerations apply to small companies that largely forfeit the use of specialized tools in the field of technical communication or where such networking is not possible, and also to medium-sized and large companies that implement so-called niche solutions for their documentation that do not have an interface to any other system in the company.

One of the resulting problems is that it is not always clear where inside the company certain processes are to be triggered and monitored. An example is the management and planning of translation jobs in medium-sized companies. Most content management systems used in this area include project management functions that are limited to finding and adding content. Additional tasks cannot be managed. As a result, the developers of TMS have increasingly started to integrate project management functions into their solutions in recent years. However, the question is whether the TMS is the correct interface for project management. Figure 6.4 shows that the translation job and thus core information are generated by means of the CMS. Accordingly, any translations that are required should also be managed by the CMS. Entering and managing this core information again at the interface from the CMS to the TMS makes this interface an error-prone break in the workflow.

For a company, the product safety and customer satisfaction are even more critical than the efficiency of the technical communication. An example is the use of synonyms. Some authors may prefer to use "port" instead of "socket" or vice versa to describe the same part of the machine. In this case translators are forced to decide whether to use only one translation or to find a synonym in their language. Usually a translator would decide to use two different terms in order to remain consistent with the original. This, in turn, leads to problems in the translation if the synonym is already used for another machine part. This phenomenon can easily be observed at companies that have been maintaining a terminology database in several languages over an extended period. The synonyms are easier to find if terminology has been recorded, as these lists or databases can be checked. It can be assumed that the problem is even larger if no official

Project management with CMS and TMS (simplified presentation)

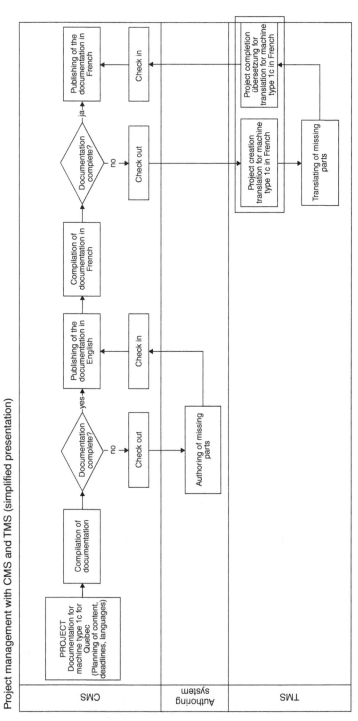

FIGURE 6.4. Typical project management workflow with CMS and TMS. These are the most complex and thorough translation processes, but they are still error-prone and may require extensive writer intervention.

terminology exists. However, the effort required to investigate this would be very high, as all existing documentation would have to be examined. But even when professional programs are used and the persons involved in creating the terminology are well trained, synonyms occur.

Synonyms are considered desirable in most languages, as repeating the same words are frequently perceived as "inelegant." However, when synonyms are used in technical communication, it is not always clear whether the synonym means the exact same thing or only something similar. This is further complicated by the fact that in my experience a strict synonym in technical documentation does not exist. As this problem already exists within the technical documentation department and in spite of the use of appropriate software, the divergence between texts written in different departments is even larger. However, texts for technical communication have to be as unambiguous as possible and must leave as little room for interpretation as possible. The ambiguity of a text can be tested by translating it and then translating the result back into the original language. The greater the similarity between the retranslated text and the original text is, the clearer and more understandable the original text is.

Remember, customers are usually confronted with several types of text written within the company. In the best case, the customer can recognize that the inconsistent text is simply a synonymous description of the same fact or object. In the worst case, the customer considers it to be a description of a different machine part or process. Furthermore, the communication between the company and the customer is affected if the names of parts or the functions of the product are not clear. If, based on the confirmation/disconfirmation paradigm, customer satisfaction is understood to be the difference between customer expectations and fulfillment of needs, such inconsistencies within and across different texts can only lead to dissatisfaction on the customer's side. No generally available studies have been done regarding the extent to which this type of dissatisfaction affects the overall satisfaction of the customer. However, it can be assumed that there is a correlation between the complexity of the product and the degree of satisfaction. The more complex the product, the more the user (in this case the customer) is dependent on an unambiguous description across all information material (usage instructions, transport and maintenance manuals, emergency procedures, spare part lists, ingredients, precautions, help files, marketing brochures, etc.). Accordingly, the customer dissatisfaction will be rampant if poor documentation accompanies the product. On the other hand, the documentation plays a less important role in the case of products that can be used easily and intuitively. Here even very poor documentation would hardly influence the customer satisfaction.

6.9   Outlook

The discussions above show that the real problem cannot be solved by using as many programs as possible. However, this does not mean that automation of the technical communication can be forfeited. Rather, it is time that the technical communication is viewed as part of the product and not as a more-or-less annoying add-on. If the technical communication is viewed as part of the product, the required software will also be integrated into the company as a whole. Unnecessary interfaces and redundancies will vanish "automatically."

At present, I don't see any solution that covers all problems and integrates all departments of a company. Much will depend upon how companies organize themselves in the future and how their self-concept will be. From my point of view, as a professional translation provider, it could have a great benefit if linguists would be involved in the research and development processes. For example, if a translator cannot find a proper designation for a component or device, it may be an indication that the design of the part itself does not match exactly the needs of the product. This implies that software tools should accompany the whole process with as few interfaces as possible.

At the moment, I would suggest that companies integrate the documentation department between writing and translating. This integration could be achieved by choosing only one planning tool for authoring and translating, which preferably should be integrated into a CMS. For smaller companies without a CMS, a small tool that tracks data flows could be sufficient.

In my judgment, the most complex or most expensive software will not guarantee the most benefits. Software must be chosen to fit existing processes if that software is to make a positive difference.

References

1. H. Bidgoli, *The Internet Encyclopedia*, Volume 1. New York: John Wiley & Sons, Inc., 2004.

2. B. Worthen, Nestlé's Enterprise Resource Planning (ERP) Odyssey. CIO. May 2002. Available: http://www.cio.com/article/31066/Nestl_eacute_s_Enterprise_Resource_Planning_ERP_Ody ssey?page=1&taxonomyId=3009, Zugriff 18.11.2011. Accessed on: March 18, 2015.

3. R. Snyder and B. Hamdan, "ERP and success factors," in *Proceedings of the ASBBS Annual Conf., Las Vegas, 2010*, vol. 17, no. 1, pp. 828–832.

4. European Agency for Health and Safety at Work, *Directive 98/37/EC—Machinery*, June 1998. Available: https://osha.europa.eu/en/legislation/directives/workplaces-equipment-signs -personal-protective-equipment/osh-related-aspects/directive-98-37-ec-of-the-european-parli ament-and-of-the-council. Accessed on: March 19, 2015.

5. International Standards Organization, *Language Codes ISO-639*, January 2001. Available: http://www.iso.org/iso/home/standards/language_codes.htm. Accessed on: March 19, 2015.

7

Communicating Food Through Muckraking: Ethics, Food Engineering, and Culinary Realism

Kathryn C. Dolan

Worst of any, however, were the fertilizer men, and those who served in the cooking rooms. These people could not be shown to the visitor,—for the odor of a fertilizer man would scare any ordinary visitor at a hundred yards, and as for the other men, who worked in tank rooms full of steam, and in some of which there were open vats near the level of the floor, their peculiar trouble was that they fell into the vats; and when they were fished out, there was never enough of them left to be worth exhibiting,—sometimes they would be overlooked for days, till all but the bones of them had gone out to the world as Durham's Pure Leaf Lard!

—Upton Sinclair, *The Jungle* [1]

Read the ingredients on the label of any processed food and, provided you know the chemical names it travels under, corn is what you will find. For modified or unmodified starch, for glucose syrup and maltodextrin, for crystalline fructose and ascorbic acid, for lecithin and dextrose, lactic acid and lysine, for maltose and HFCS [high fructose corn syrup], for MSG and polyols, for the caramel color and xanthan gum, read: corn. Corn is in the coffee whitener and Cheez Whiz, the frozen yogurt and TV dinner, the canned fruit and ketchup and candies, the soups and snacks and cake mixes, the frosting and gravy and frozen waffles,

Communication Practices in Engineering, Manufacturing, and Research for Food and Water Safety, First Edition.
Edited by David Wright.

the syrups and hot sauces, the mayonnaise and mustard, the hot dogs and the bologna, the margarine and shortening, the salad dressings and the relishes and even the vitamins.

—Michael Pollan, *The Omnivore's Dilemma* [2]

7.1 Muckraking and Promoting Food Safety

Historically, one of the most effective forms of communicating food safety concerns to the public has been through the genre of muckraking literature. The authors who employ this technique incorporate realism and sentimentalism into their work in order to educate society and potentially create lasting reform. In this chapter's epigraphs, two muckraking authors use descriptive language to educate their readership about significant food production hazards of their times. The first passage above is one of the most memorable ones in a book famous for describing the horrors of the Chicago meat-packing industry during the early twentieth century: *The Jungle* [1]. In this scene, in addition to a description of the extreme odor of fertilizer workers, Upton Sinclair explains in gruesome detail that there may be cannibalistic implications in the lard that consumers eat. Indeed, cannibalism is beyond a "peculiar trouble." This extreme example concludes a list of horrors that affect both the workers victimized by the Chicago Beef Trust as well as the consumers unwittingly eating an unknown quantity of hazardous elements under the label of "Pure." Sinclair's exclamation point is warranted in this instance. "Durham's Pure Leaf Lard" becomes an ironic name, and his exclamation point emphasizes this irony. Grisly accounts like these—including but not ending with cannibalism—were powerful and effective.

In the second passage above, from a twenty-first century example of food safety themes in literature, *The Omnivore's Dilemma* [2], Michael Pollan similarly uses shock value to educate his audience about the foods they eat—specifically the corn sweetener called high fructose corn syrup (HFCS). In this long list, he enumerates a wide, seemingly exhaustive, list of corn additives in foodstuffs—all of which have been disguised in some fashion. He includes obviously unsafe food items like "MSG," alongside the more seemingly safe "vitamins" in order to make the list truly exhaustive. Moreover, foodstuffs that might seem "pure" are once again implicated in their hidden ingredients—in this instance the maple syrup.

The foods made for public consumption are shown to contain previously unknown and potentially hazardous elements in both epigraphs, and food safety is a primary concern for authors who use the muckraking genre—in novel and non-fiction form—to give readers examples of key safety issues during two distinct periods in the history of food production and engineering in order to force consumers to reconsider what they know about the foods they eat.

Upton Sinclair's *The Jungle* remains perhaps the most influential work of US literature in terms of influencing genuine change in food safety procedures, in large part due to his graphic descriptions of tainted beef. Sinclair's account of the great "aggregation of labor and capital" managed ultimately to bring about changes in regulation [1, p. 51]. The Pure Food and Drug Act and Meat Inspection Act of 1906 quickly followed the publication of his book, the former being the precursor to the modern Food and Drug Administration (FDA) [3]. A century later, Michael Pollan examines prevalence of high fructose corn syrup in the US food supply in his modern-day muckraking works *The Omnivore's Dilemma* and *In Defense of Food* [4]. Pollan argues that food processing taken to the extreme in the twenty-first century has become a modern version of Sinclair's Chicago Beef Trust. For Pollan, the corporations responsible for the failure to educate the public about food safety issues are the corn processors ADM and Cargill. Both authors use elements of realism and sensationalism to communicate with readers at the rational and emotional levels. This combination gives literature the ability to be a powerful means of educating an audience on key food security concerns and of bringing about positive change. Sinclair and Pollan therefore represent a particularly effective form of written communication that rakes up the muck of the food industry—specifically looking at issues of food processing—to allow consumers to make healthier choices regarding their food. They engage with the ethics of food manufacturing through their literature in productive ways.

7.2 Culinary Realism and Food Safety

In an age of multimedia alternatives, literature continues to communicate with a broad audience through an internalized empathetic experience through the reader's imagination. Jason Pickavance [5, p. 91] names the particular style of writing used in Sinclair, and arguably with a postmodern twist in Pollan, as "culinary realism." This genre occurs where "literary realism meets the rhetoric of food and diet reform." In this reading, the attention to detail that is a foundation of the genre of literary realism applies to descriptions of food production. Authors apply culinary realism to their work by focusing on the materiality of food items in order to make a believable reading experience. They frame the discourse, and this gives their claims more substance. While their focus is often on the relationship between workers and consumers and the food industry, this literary technique can be usefully applied in terms of food engineering companies, as well. Peter Modin and Sven Hansen [6, p. 314] list "seven practical principals" of describing food safety and risk to consumers as part of an ethics of food production. They recommend that food companies

1. Be honest and open
2. Disclose their own interests in the matter
3. Describe any and all knowledge obtained
4. Quantify risks if possible
5. Describe uncertainties
6. Take the public's concerns into account
7. Take the rights of others seriously.

Culinary realism in muckraking literature thus can be used to serve two functions. It educates consumers while encouraging food companies to follow seven delineated principles of ethical food production, and reports any companies that fail to follow some level of ethics.

Returning to the twinned epigraphs, Sinclair applies culinary realism in his work. He tells his readers exactly what they are eating and preparing at home. In this technique, he is demonstrably effective. Sinclair moves away from his usual intimate detail regarding his main characters, the Rudkuses, to discuss the nameless workers. He goes farther with his direct address, as well. He names his consumers the same people as would be the "visitors" prevented from seeing the "exhibition" of the workers within the packing houses. They are denied access to information that he can provide them, to great effect. His work then becomes a public service. Similarly, Pollan speaks to the "consumer" directly. He generalizes his descriptions for effect. Furthermore, there is an element of bringing something hidden into the open in both authors' works. While Sinclair describes the horrors inside a meat-packing plant, Pollan takes this opportunity to speak to his twenty-first-century readers about HFCS in an astounding array of everyday items, what consumers should know about the sweetener, and what they can do with that knowledge.

Muckrakers use the tools available through culinary realism in literature to communicate some of the major food safety issues of their time. In a 1906 article for *Cosmopolitan*, Sinclair [7, p. 593] writes that he "spared no pains to get every detail exact." This realistic detail made his work particularly effective at communicating food risks to the consumers of the early twentieth century. Moreover, he cites his influences as Harriet Beecher Stowe's *Uncle Tom's Cabin* [8], another US classic of literature resulting in social change. Like Stowe, Sinclair draws attention to a systemic ethical issue through the form of the novel. This need for muckrakers has continued into our twenty-first-century culinary jungle. The tradition continues with Pollan, a journalist whose goal is to educate the public about food safety in the case of modern processing and additives in foodstuffs. In either case, it seems, as Pickavance [5, p. 91] has observed,

Realism mattered most when it came to the stomach.

More than other consumables, issues of food safety are naturally significant because it is in food's nature to be incorporated into our bodies; therefore our concerns regarding food are necessarily more intense than other issues. Sinclair and Pollan [2, p. 86] use this communication technique to draw attention to organizations like the Chicago Beef Trust and Cargill and ADM, "the two companies who wet-mill most of America's corn," respectively, looking specifically at poisoned meat and corn sweeteners to do so.

7.2.1 Tubercular Beef in *The Jungle*

The impact of *The Jungle* [1], Sinclair's 1906 muckraking classic, has been nearly unparalleled in US literature and history and is the foundation for many established food safety regulations in the United States. It shows an example of the dramatic real-world results possible through effective communication. Any work on literature and food or public policy must engage with *The Jungle* and how it has defined the field. Audiences

avidly followed the plight of the Rudkuses when the story was first published serially in
Appeal to Reason [9], between February 25 and November 4, 1905, and even more so
as a novel—the novel became an international bestseller within weeks of publication. In
fact, its impact is perhaps the most important aspect of the novel, and this is perhaps the
true reason that *The Jungle* is taught regularly in college undergraduate classes in US
literature and history.

> If authors hope to influence others, ideally to create positive social change, Sinclair's
> work remains an ideal model.

Sinclair uses a horrific culinary realism as the most effective way to communicate
food safety issues to his audience.

For those unfamiliar with Sinclair's novel, it follows the plight of Jurgis Rudkus
and his family as they come to grisly and tragic ends at the hands of the Beef Trust
in Chicago's meat-packing industry during the early twentieth century. Historically,
advances in refrigeration and the expansion of the railroads, combined with the indus-
trialization of the meat-packing industry, changed the way people consumed food over
the course of the nineteenth century. This made Midwest cities like Chicago wealthy as
they carved a niche for themselves in US food production by processing the livestock
raised on Western ranches to sell to the markets of the East. William Cronon [10, p. 223]
observes that as the livestock industry in the West exploded between 1860 and 1890,
the "steep growth of Chicago's beef packing" correspondingly began in earnest in the
1870s. By the turn of the century, Chicago dominated the nation's production of beef
and pork. The "Beef Trust," as the powers behind the industry were known, vertically
controlled the transportation, stockyards, and retail portions of the meat-packing industry. Due to this level of control of the entire industry, the Beef Trust was to some extent
impervious to any system of checks and balances. Even before Sinclair researched *The
Jungle*, Ronald Gottesman writes that [11, p. xv], *The New York Journal* had published
a report exposing the common practice of packing condemned meat in 1899.

Meanwhile, Theodore Roosevelt had tried to draw attention to the unsafe meat
being fed to US military personnel in Cuba during the Spanish–American War. However,
Roosevelt's arguments were unsuccessful in creating any reform at that time. Not only
was the Beef Trust sending unsafe meat to consumers, including soldiers on the front
lines, it was also economically damaging small-scale butchers who were its primary
competition. The Beef Trust practically created a new industry out of processing the
by-products of the beef and pork from their packinghouses. While a traditional butcher
would not have the means to process small amounts of the animal after having been
dressed for beef, the enormous size of Chicago's packinghouses made the practice
profitable. In fact, the Beef Trust was able to augment their profits from this industry
in order to be able to undercut smaller butchers on the price of their dressed beef,
eventually driving them out of business entirely. The favorite method of processing
these by-products was in bologna sausages, as well, as they could "hide such a multitude
of sins" [10, p. 252].

Once more, Sinclair's culinary realism is useful to show scenes of hazardous food production practices, specifically in sausage engineering:

> Cut up by the two-thousand-revolutions-a-minute flyers, and mixed with half a ton of other meat, no odor that ever was in a ham could make any difference. There was never the least attention paid to what was cut up for sausage; there would come all the way back from Europe old sausage that had been rejected, and that was moldy and white—it would be dosed with borax and glycerine, and dumped into the hoppers, and made over again for home consumption. There would be meat that had tumbled out on the floor, in the dirt and sawdust, where the workers had tramped and spit uncounted billions of consumption germs. There would be meat stored in great piles in rooms; and the water from leaky roofs would drip over it, and thousands of rats would race about on it. It was too dark in these storage places to see well, but a man could run his hand over these piles of meat and sweep off handfuls of the dried dung of rats. These rats were nuisances, and the packers would put poisoned bread out for them; they would die, and then rats, bread, and meat would go into the hoppers together. This is no fairy story and no joke; the meat would be shoveled into carts, and the man who did the shoveling would not trouble to lift out a rat even when he saw one—there were things that went into the sausage in comparison with which a poisoned rat was a tidbit. There was no place for the men to wash their hands before they ate their dinner, and so they made a practice of washing them in the water that was to be ladled into the sausage. [1, p. 163–164]

If the scene from the epigraph were the most striking description of the horrors of the meat-packing industry in Chicago, this second selection comes close to meeting and perhaps exceeding the first in terms of specific food safety issues. Sinclair shows the extreme lack of concern for customer safety on behalf of the Beef Trust; indeed, consumers are being subjected to poisoned foods. Not only are the poisoned rats and bread described as being processed into the sausage, but also Sinclair compounds this by alluding to worse elements in the sausages, the likes of which make such poisons "a tidbit" [1, p. 164]. This comment alludes to the quote in the epigraph, where consumers are introduced to the extreme of unwitting cannibalism in their sausages. There is no part of this process that is designed with the safety of the consumer in mind—it is all profit-based. Thus, people who purchase this kind of food would be right to be alarmed to discover what truly is in their morning sausages.

In order to effectively communicate his point, Sinclair incorporates several sensory descriptors into the above passage, for example, terms like "odor," "spoiled," and "white and moldy." His readers can visualize and to some extent hear workers spitting, washing their hands, and tramping through the work floor in this scene. Moreover, nothing is in moderation in this passage. Sinclair uses large quantifiers like "two-thousand-revolutions-per-minute," "half a ton," "uncounted billions," "thousands of rats," and "handfuls of dried dung" throughout. These numbers invoke a sublime reaction; they are simply too big to be comprehended. Sinclair's food safety claims are extreme and may have strained the credulity of his readers, even though they are researched and factual. Therefore, part of his skill as a communicator is to reassure his readers of the truth of his statements, though his truth is stranger than the fictions of which they might be familiar. To maintain his audience, midway through the litany of abuses, Sinclair pauses and notes that *The Jungle* is neither a "fairy story" nor a "joke." This insertion adds a somber element to the rest of his images. A scene like this would lead Sinclair's readers to perhaps disbelieve his statements, but he reminds his readers to trust him as they cannot trust their food producers.

Sinclair [1, p. 116] then shows the corruption involved in meat-packing companies and how their unethical practices become further food safety issues for consumers. An official, after noticing tubercular steer carcasses "which therefore contained ptomaines, which are deadly poisons, were left upon an open platform and carted away to be sold in the city; and so he insisted that these carcasses be treated with an injection of kerosene—and was ordered to resign the same week!" The business interests who profited from the cheap—if unhealthy—meat were more concerned with maintaining their profits than in complying with established food safety laws. Furthermore, they had the power to control the distribution of information and enforcement of food safety laws. According to Sinclair [1, p. 116] the Beef Trust took the matter a further step, forcing Chicago's mayor to "abolish the whole bureau of inspection." In a passage like this, Sinclair shows that the Beef Trust focuses only on profits, to the detriment of their customers. Understandably, the consumers who read Sinclair's accounts found this an unacceptable abuse of their trust and forced changes that have remained into the next century.

President Theodore Roosevelt was one such consumer. After reading Sinclair's culinary realist account of the Chicago Beef Trust, he ordered a committee to research the facts regarding Chicago's meat-packing industry, which were then described in a Congressional Report in the months following the publication of *The Jungle*. James Bronson Reynolds and Charles P. Neill, the key investigators of the report [12], noted horrible working conditions among the stockyards of Chicago's Beef Trust. Figure 7.1 shows a 1900 image of a meat-packing plant. They supported the worst of Sinclair's claims: they were especially concerned about lavatories, lighting and heat in the buildings, lack of ventilation, and unsanitary meat handling. They stated, "We saw a hog that had just been killed, cleaned, washed, and started on its way to the cooling room fall from the sliding rail to a dirty wooden floor and slide part way into a filthy men's privy. It was picked up by two employees, placed upon a truck, carried into the cooling room and hung up with other carcasses, no effort being made to clean it" [12, p. 7). Moreover, they observed that "the fumes of the urine swell the sum of nauseating odors

FIGURE 7.1. The Chicago meat-packing industry. This image shows a working turn-of-the-century Chicago meat-packing site. In 1900 this was a completely unsafe environment for both workers and consumers. See the Library of Congress full image at http://www.loc.gov/item/2007663982/

arising from the dirty-blood-soaked, rotting wooden floors" make for "fruitful culture beds for the disease germs of men and animals." Furthermore, they blamed the Beef Trust executives for the conditions directly, noting that "expense" is "so often heard" as the only excuse given—or apparently required—for the various unsafe procedures [12, p. 5]. The concern for economy and profits that hurt small-scale butchers is taken to an extreme in this instance in terms of a failure in food safety and communication.

Sinclair uses culinary realism to describe hazardous conditions in Chicago's food production industry. In his pedagogical analysis of *The Jungle*, Christopher Phelps writes that it remains one of the top five supplementary texts assigned in US historical surveys [13]. In addition, it is commonly taught in US literature survey courses (see Figure 7.1). The novel has gone from muckraking tract to a canonical work. This novel represents "exposé journalism that blends revealed fact with moral indignation in the pursuit of social reform" [11, p. 4]. Part of that success is due to its purpose of educating the public on what they eat. This does not mean, of course, that Sinclair's novel is a flawless work of fiction or even that the impact the novel had was the one Sinclair intended. He famously complained at one point that [4, p. 594] "entirely by chance I had stumbled on another discovery—what they were doing to the meat-supply of the civilized world ... I aimed at the public's heart, and by accident I hit it in the stomach." However, these elements do not lessen the force of the novel; rather, they demonstrate the inherent significance of food as a subject. It is the nature of food to be ingested into the body, giving it an almost mystical quality and the potential to be an object of fear.

The fact that the majority of US consumers are not directly connected to the chain of production of their food means that they must take their relationship with food and its producers a great deal on faith. Sometimes, that faith can be misplaced or abused. Pickavance [5] argues that Sinclair does, in fact, mean to focus on food safety issues throughout the novel. They are intended to be there as a necessary component of the horrors of the Chicago meat-packing industry.

Sinclair spends the majority of his novel describing the manner in which the US industrial agriculture seen in the Beef Trust is destroying its workers—primarily recent immigrants—within the stockyards of Chicago's meat-packing industry. Later in the novel, though, he suggests potential reforms to help workers and consumers. He [1, p. 407] imagines "apples and oranges picked by machinery, cows milked by electricity" in a kind of pre-Green Revolution scenario "by which the population of the whole globe could be supported on the soil now cultivated in the United States alone!" While these images sound like a kind of food engineering science fiction, they demonstrate Sinclair's conscious effort to find a remedy to the issue of the abuses of a labor class through a reform of US agriculture via advances in science and technology.

At the turn of the twenty-first century, the issues involved in food safety return to Sinclair's image of an overly engineered monoculture.

Not surprisingly, then, a new generation of muckrakers have come about to challenge the new trusts—in this case represented by the corporations of ADM and Cargill—as they navigate the brave new world of food engineering. Authors like Michael Pollan work to educate a twenty-first-century audience on their food consumption, one that has changed a great deal since Sinclair's food safety novel, while remaining similarly disturbing.

7.3 High Fructose Corn Syrup in *The Omnivore's Dilemma and In Defense of Food*

In the muckraking tradition applied so successfully by Upton Sinclair, Michael Pollan once again shows that literature can and should challenge US businesses in issues of food safety by communicating directly to consumers. A century after Sinclair's fiction *The Jungle*, Pollan's nonfiction work *Omnivore's Dilemma* [2] and his follow-up *In Defense of Food* [4] once again expose potential dangers in the US food industry in an engaging narrative manner using culinary realism. Pollan creates a modern classic that asks many of the same questions as the classic muckraking works of a century earlier. He also focuses once more on the producers and their concern for profits as opposed to the workers who cannot seem to survive within the industry. In this modern food industry, the small-scale corn farmers and producers are being harmed when faced against conglomerations like AGM and Cargill, new iterations of the Chicago Beef Trust. In the twenty-first century, Pollan focuses specifically on the potential dangers of HFCS and its omnipresence in the US diet. His work has proven immensely successful, and the debate between sugar and HFCS rages on into the 2010s—even involving the FDA and bringing the discussion back to the organization Sinclair helped to form at the turn of the previous century. Pollan continues Sinclair's legacy of furthering a food safety conversation regarding alternatives to intensive food processing, challenging consumers to educate themselves and bring modern food engineering out of the laboratory and back to the earth.

While Pollan's writing discusses a variety of industrialized food products, his primary target is HFCS. Corn sweeteners are the most dangerous of the food-like substances being added to US foods, in his opinion. He argues that the prevalence of cheap calories—primarily in the form of HFCS—is a primary reason for the obesity epidemic, the rise in diabetes II, and other health problems that affect a billion people worldwide. The history of HFCS has been progressing in the United States since the nineteenth century. In the first half of the nineteenth century, surplus corn became animal feed or alcohol. Those were its primary alternative forms. However, corn was soon broken into component parts such as glucose by 1866, and corn syrup became the earliest domestic cane sugar replacement [2, p. 88]. This is a technological advance that has had lasting consequences. Corn syrup would become the dominant sweetener over the next century, as technologies worked to refine corn syrup into its newest permutation, HFCS. Pollan states [2, p. 89] that high-fructose corn syrup—a fructose/glucose blend—came onto the market in the 1970s. At the time he writes it accounts for 530 million bushels of corn annually. Pollan has given his readers a basic history of the development of this particular sweetener, and some of the consequences—particular regarding health—of that development. Interestingly, while corn-based sweeteners have been used since the nineteenth century, it is the emphasis on HFCS that seems to have truly affected consumer health. Pollan, therefore, focuses on that particular corn sweetener through his work.

Omnivore's Dilemma is a work of journalism that deliberately attempts to understand the modern industrial food world in which US consumers live in the twenty-first century. Divided into three sections, the longest is the industrial diet section, in which HFCS plays a dominant part. Pollan travels to Iowa to investigate a modern corn farm, researches the subsidized and industrialized aspects of current methods of farming corn, and attempts to witness the processing of corn into a medley of substituent parts that are then incorporated into almost every aspect of the modern US diet. This wet-milling process fascinates him, and he works to understand what exactly occurs [2, p. 90].

> Step back for a moment and behold this great, intricately piped stainless steel beast [corn processor]: This is the supremely adapted creature that has evolved to help eat the vast surplus biomass coming off America's farms, efficiently digesting the millions of bushels of corn fed to it each day by the trainload. Go around back of the beast and you'll see a hundred different spigots, large and small, filling tanker cars of other trains with HFCS, ethanol, syrups, starches, and food additives of every description. The question now is, Who or what (besides our cars) is going to consume and digest all this freshly fractionated biomass—the sugar and starches, the alcohols and acids, the emulsifiers and stabilizers and viscosity-control agents? This is where we come in. It takes a certain kind of eater—an industrial eater—to consume these fractions of corn, and we are, or have evolved into, *that* supremely adapted creature: the eater of processed food.

Pollan gives a detailed description of the beast-like machine that processes corn into a hundred smaller components. He lists "HFCS, ethanol, syrups, starches, and food additives" as part of the exhaustive list of foodstuffs that have become corn-enriched through the technology exemplified through the wet-milling process of ADM and Cargill, primarily. Most significantly, he concludes with an image of a corresponding machine-like beast who is seemingly capable of consuming all that corn in its myriad products: the modern-day US consumer, "the eater of processed food," and "an industrial eater." People are no longer consumers of food in an idyllic, almost nostalgic manner.

It takes a certain kind of eater—an industrial eater—to consume these fractions of corn.

US consumers have become machine-like in their ability to consume as many calories and industrial food products as are being produced. This becomes a nightmarish scene for Pollan.

In this way, Pollan's attack against HFCS becomes highly focused. Pollan argues [2, p. 104], "Since 1985, an American's annual consumption of HFCS has gone from forty-five pounds to sixty-six pounds." Meanwhile, US consumption of sugar has not declined: "During the same period our consumption of refined sugar actually went up by five pounds. What this means is that we are eating and drinking all that high-fructose corn syrup *on top* of the sugars we were already consuming." In keeping with his earlier metaphor of a machine-like consumer of calories—especially HFCS calories—as opposed to a human eater of foods, Pollan observes that US consumers have dutifully increased their intake of all sweeteners by 30 pounds per person annually. People are eating what they are being told to eat—hungry or not, healthy or not, safe or not. In this way, US eaters successfully consume the corn surplus with a machine-like efficiency. HFCS is everywhere. "Read the food labels in your kitchen and you'll find that HFCS has insinuated itself into every corner of the pantry…into the ketchup and mustard, the breads and cereals, the relishes and crackers, the hot dogs and hams" [2, p.104]. Pollan's point here is that the ubiquity of HFCS in the US diet is mostly hidden. People do not realize that they have added such a large amount of sweeteners to their diets. Industrialized foods are not like actual foods—they are filled with preservatives, fillers, and HFCS. These are ingredients and elements that consumers would not willingly add to their own food, but eat without complaint in a wide variety of other products.

Pollan continues his attack against HFCS in his follow-up work, *In Defense of Food*. At the end of his more pragmatic work, he gives readers a list of instructions in terms of eating healthier and choosing safer foods within the grocery store. Indeed, his first line reads, "Eat food. Not too much. Mostly plants" [4, p.1). This book is in itself a sign of the results of *Omnivore's Dilemma*. Pollan found himself confronted by a generation of eaters who had become terrified of the overly processed foods, specifically HFCS, in their diets; however, these same consumers had no idea how to improve their diets. His later work gives active suggestions to his readers of purchasing foods from the periphery of the grocery store—avoiding the boxes in the aisles where the problems of the highly processed foods are more intensified. The periphery, he reasons, is the region where actual foods reside: produce rather than boxed or canned vegetables and fruits, deli meats ordered from a person behind a counter rather than prepackaged meats, and dairy and eggs as opposed to heavily processed flavored sweet yoghurts and other items. One of his direct recommendations states [4, p. 150], "Avoid food products containing ingredients that are a) unfamiliar, b) unpronounceable, c) more than five in number, or that included) high-fructose corn syrup." In this entertainingly simple and straightforward list of food rules, HFCS is the only ingredient that is directly targeted as being necessary to avoid. His rationale is that HFCS is so prevalent in our diets that we should make every effort to reduce our intake of it whenever possible—exerting any amount of control that we actually possess over what we consume.

Consumers were delighted and relieved with this straightforward advice—even in the case of something as seemingly strict as "avoid high-fructose corn syrup." In fact, following the success of Pollan's pair of culinary realist texts, consumers to a large extent did exactly as Pollan asked them to do—they avoided HFCS. Additionally, shoppers successfully followed much of his advice in addition to this specific admonition. They shopped on the perimeter of the grocery store—where the produce and deli are usually located. They bought foods with fewer overall ingredients, particularly working to eliminate some of the more unpronounceable items. People loved to think about whether or not their grandmothers would have recognized certain food products being sold as foods. All of Pollan's suggestions made an impact on consumers, but none did so more than his admonition against HFCS. There, shoppers were inflexible. This consumer pressure exerted an impressive and demonstrable influence on the industry.

As consumers began to vote with their wallets, or as Pollan argues, began to vote with their forks, food companies responded to their changing demands. Products started claiming to use cane sugar as opposed to HFCS. A 2009 Consumer Reports [14] article notes the Coca-Cola Company marketed the Mexican-made cane sugar version of their staple cola in US stores. The Sara Lee bread company also advertised a "No High Fructose Corn Syrup" version of their bread in 2010 [15]. In addition, documentaries were made that also took on the issue of HFCS, such as *King Corn* [16]. In short, there was a publicity nightmare around HFCS. It became the demonized food item, and consumers were determined to do something about it. According to the Sugar Association website, the sales of foods containing HFCS dropped 11% from 2003 through 2008 [17]. The public outcry against HFCS has been intense since the 2006 publication of *Omnivore's Dilemma* and 2008 publication of *In Defense of Food*, and it has remained a primary issue in shopping for food well into the 2010s.

Beginning in 2008, therefore, the corn industry has been aggressively responding to this intense demonstration of public opinion against their most profitable product. In defense, the Corn Refiners Association—made up of the wet-milling operations of ADM and Cargill discussed in Pollan, among others—responded with an advertising campaign in which they claimed HFCS as being the same as sugar. Through a series of print and television advertisements, actors have conversations in which their alarmism against HFCS is alleviated when they are reassured that "sugar is sugar." For an example of the print campaign, see Figure 7.2. These ads had mixed results, and the Corn Refiners Association did not stop there. The association followed this act by filing a petition with the FDA in 2010 to change the name of HFCS to "corn sugar," hoping to circumvent the negative publicity in that way while maintaining the profitability of inserting HFCS in almost all food items in the grocery store.

What followed was a publicity storm, with experts on both sides of the HFCS debate weighing in. Ultimately, in 2012, the FDA rejected the appeal, and the name remains high fructose corn syrup into the 2010s [18]. In 2011, the Sugar Association filed a lawsuit against the Corn Refiners Association for deliberately misleading the public. They maintain that HFCS is not corn sugar [19]. Unsurprisingly, the Corn Refiners Association countersued in 2012 [20]. This issue is ongoing into the 2010s.

The underlying issue here is food safety, and a demand for an increased level of transparency in the companies producing the foods being consumed, a version of the ethical chart listed above. Paraphrased, the key issues for companies concerned are honesty, full disclosure of interest, knowledge, risks, and uncertainties, as well as the need to act in consideration of the public's wants and needs. Ideally, the winners in firestorm of media, news, and lawsuits will be the consumers. Foods will become healthier and safer because of this intense level of scrutiny. Pollan's incisive reports helped to start this movement, and he remains the poster bearer of healthy foods and "No HFCS" well into the twenty-first century.

Michael Pollan is one of the most recent in a long line of muckrakers—journalists, novelists, and documentary filmmakers—who work to bring issues of public safety to light for the sake of consumers. Pollan continues his work through his website,

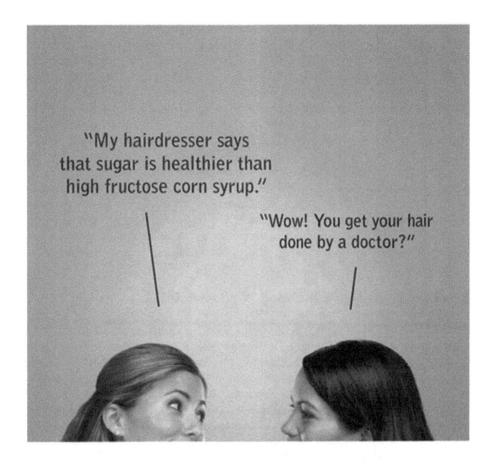

FIGURE 7.2. Corn Refiners Hairdresser Ad. This image is an example of the Corn Refiners Association advertising campaign in 2002, designed to battle the backlash against HFCS through irony [21].

michaelpollan.com, where he links food-related articles from other news sources. While Pollan's message might seem to be targeted to a different group of consumers than Sinclair's workers in *The Jungle*, both authors have fundamentally universal underlying concerns. Sinclair's concern about tubercular beef touched a national nerve. Meanwhile, Pollan's concern about the obesity epidemic, and the fear that today's children might have shorter life expectancy than their parents—a first in human history—based largely on their diets of highly processed foods, is an enormous issue that goes beyond class. In both cases, the food safety issues are extreme, and so a book like Pollan's has a similarly extreme effect on society. His culinary realism has caused a chain of reactions that resonates into the 2010s. Consumers have been, to some extent, scared straight. A perhaps unforeseen consequence of the debate between HFCS and sugar is the development of yet another sweetener, Stevia. A new sugar alternative, it does not yet have the negative press of all previous artificial sweeteners, sugar, or HFCS. For now, perhaps, Stevia has become the sweetener of choice for a nation addicted to sugar but increasingly concerned with the consequences of that national sweet tooth. See Figure 7.2.

7.4   Literature as a Watchdog in Food Safety

Reviewing one of the fictional classics in muckraking alongside a modern nonfiction work provides a useful comparison of methods of successfully communicating issues in food safety and engineering. These texts provide readers a way to empathetically engage with some of the vital concerns in terms of the foods they eat. In this way, culinary realism provides an invaluable resource to the consumer; they scrutinize food producers and call for transparency in issues of food production. They serve a necessary communicative role and will continue to do so as long as people are distanced from the source of their food. Sinclair wrote of the horrible working conditions and tainted meat in the early twentieth century. Pollan analyzed the harmful health effects and omnipresence of high fructose corn syrup in the early twenty-first century. Their lessons on transparency, the importance of remembering the human end user who possesses the rights to health and well-being regarding food is a necessary one for food engineers and their corporations, as well.

Muckrakers become a kind of watchdog element that promotes a level of transparency between corporations and consumers. These communicators serve a necessary role in providing lessons on food safety to the public. This is beneficial for food safety engineers, as well, whose ultimate goal must be the safety of consumers who use their food products.

Muckraking texts provide food engineers with what can be seen as the equivalent of a lesson in ethics.

These works should perhaps become required literature for engineers in this field. Corporations like Cargill or ADM can use this literature, as well, to make themselves as focused on food safety as they can be. As with the sciences, food engineering concerning food safety must be held to the strictest standards—for the benefit of the people who are the ultimate consumers of their food production. This is the purpose for muckrakers—historical and modern—and this becomes particularly relevant when the consumable is as literally incorporated into a consumer's body as is the case with food production.

According to Ulrich Beck [22], we are currently in a "risk society." The original muckrakers of the nineteenth century worked to teach the public that their trust was often misplaced in areas like food safety. Since that time, though, connection to food production on the individual level has greatly lessened, to the point where in the twenty-first century it is common for people to be wholly unfamiliar with the means of the production of their food. Indeed, recalls and food-related safety scares have become practically commonplace in recent history. Concerned figures have to make the issue personal again through the combination of factual and imaginative work as seen in these muckraking examples. The frequent and often severe food recalls in recent history are modern examples of the issues being dealt with in Sinclair's classic. There have been hundreds of food recalls between the years 2005 and 2010, some of the more interesting ones being a 2008 recall of tomatoes due to a *Salmonella* outbreak that concerned food trading routes between Mexico and the United States and the 2006 recall of bagged spinach due to an *E. Coli* outbreak that left grocery store shelves empty and the spinach industry devastated [23]. In the year 2012, yet another *Salmonella* outbreak was found among various Sunland products, including high-end peanut butters. The initial announcement about the need for a recall was made on September 22, 2012, and by October 13, 2012, the FDA had confirmed the presence of *Salmonella* in the food product. By the end of October, over three dozen people had confirmed cases of *Salmonella* connected to tainted foods across 20 states. While none of these cases proved fatal, a disturbing theme is the fact that two-thirds of the cases were in children under 10 years of age [24]. This recent example of food safety and the need for communication regarding food safety issues once again demonstrates the importance of the work of muckrakers like Sinclair and Pollan.

One of the most frightening recent recalls—hearkening to the true issue behind muckraking works and issues of food safety—was the 2007 Menu pet food recall, simply because it made evident how unsecure our modern global food system is. As Marion Nestle 25, pp. 150–151 [25, pp. 150–151] warns, the Menu recall has implications far beyond that of poisoned pet food. She fears the recall was a "Chihuahua in the Coal Mine," as is the subtitle of her book on the topic. Ultimately, the human food chain is in fact no safer than pet food at the ingredient level. Terrorists need not use bombs to do the greatest damage to the greatest number of people in the United States; all they have to do would be to poison or taint one of the ingredients in modern processed foods. Such a maneuver would be almost impossible to trace, deadly, and widespread. In fact, following the Menu pet food recall event, the government developed the Food Protection Plan to "Prevent, Intervene, Respond" when "high-risk foods" become excessively dangerous. Among other listed responsibilities, they "[a]llow FDA to require preventative controls to prevent intentional adulteration by terrorists or criminals at points of high vulnerability in the food chain." With foods like this, a new generation of muckrakers ask, who needs enemies?

7.5 The Effects of Literature on Everyday Practices

So, what are the success rates of muckraking works in effectively communicating food safety issues to the public and potentially bringing about some form of social change? Authors like Sinclair and Pollan are complemented by the work of muckrakers like Marion Nestle and a generation of others. Sinclair's culinary realism remains a model for effective muckraking journalism. Pollan, meanwhile, reimagines muckraking for the new century. He adds a postmodern element to culinary realism, and the result is also effective for dealing with food safety hazards at a time of drastic changes in engineered food production. Ultimately, he educates his readers on their inherent power as consumers and their need to remain diligent in order to exert control over their own food safety. The results are striking—people do have the power to effect change in the twenty-first century in terms of food security. There has been an explosion of small farms, farmer's markets, and heirloom seed development. In no small degree, these have been positive changes brought about through the work of modern-day muckrakers who apply literature to educate the public about food safety issues most effectively.

References

1. U. Sinclair, *The Jungle* New York: Penguin, 1986.

2. M. Pollan, *The Omnivore's Dilemma: A Natural History of Four Meals.* New York: Penguin, 2006.

3. U.S. Department of Health & Human Services, "Federal Meat Inspection Act," May 2009. Available: http://www.fda.gov/regulatoryinformation/legislation/ucm148693.htm. Accessed on March 12, 2015.

4. M. Pollan, *In Defense of Food: An Eater's Manifesto.* New York: Penguin, 2008.

5. J. Pickavance, "Gastronomic realism: Upton Sinclair's *The Jungle*, the fight for pure food, and the magic of mastication," *Food and Foodways: Explorations in the History and Culture of Human Nourishment*, vol. 11, no. 2–3, pp. 87–112, 2003.

6. P.G. Modin and S.O. Hansson, "Moral and instrumental norms in food risk communication," *Journal of Business Ethics*, vol. 101, no. 2, pp. 313–324, 2011.

7. U. Sinclair, "What life means to me," *Cosmopolitan*, vol. 41, pp. 591–594, 1906.

8. H.B. Stowe, *Uncle Tom's Cabin.* Boston: John P. Jewett & Company, 1852.

9. U. Sinclair, "The Jungle: Chapter One," Appeal to Reason, no. 482, Feb. 25, 1905.

10. W. Cronon, *Nature's Metropolis: Chicago and the Great West.* New York: Norton, 1992.

11. R. Gottesman, "Introduction," in *The Jungle*, 25th ed., New York: Penguin, 1986, pp. xv.

12. J.B. Reynolds and C.P. Neill, Congressional Report 873, June 4, 1906.

13. C. Phelps, "How should we teach *The Jungle*?" *The Chronicle of Higher Education*, vol. 52, no. 26, pp. B10–B12, 2006.

14. Consumer Reports. (2009, Jun.). Coke vs. Coke: A tale of two sweeteners [Online]. Available: http://www.consumerreports.org/cro/magazine-archive/june-2009/food/coke-vs-coke/overview/coke-vs-coke-ov.htm. Accessed March 10, 2015.

15. E. Schroeder, "Sara Lee reformulates some bread with no HFCS," August 4, 2010 [Online]. Available: http://www.bakingbusiness.com/News/News%20Home/New%20Products/2010/8/Sara%20Lee%20launches%20reformulated%20bread%20with%20no%20HFCS.aspx?cck =1. Accessed March 10, 2015.

16. King Corn, Directed by Aaron Woolf, 2007. Greene, IA: Cinedigm Entertainment, 2008.

17. The Sugar Association, The Timeline, 2013. [Online]. Available: http://sugar.org/cra-lawsuit/the-timeline/.

18. The Sugar Association, FDA Petitions, 2013. [Online]. Available: http://sugar.org/cra-lawsuit/the-fda-petitions/.

19. The Sugar Association, The Lawsuit Against the Corn Refiners' False Advertising, 2013. [Online]. Available: http://sugar.org/cra-lawsuit/.

20. D. Knowles, Corn Refiners Counter Sue the Sugar Association, September 4, 2012. [Online]. Available: http://www.corn.org/press/newsroom/corn-refiners-counter-sue-sugar-association/.

21. Corn Refiners Association. Advertisement 2008. Available: http://www.pwrnewmedia.com/2008/cornrefiners082808/downloads/hairdresserAd.pdf.

22. U. Beck, *Risk Society: Towards a New Modernity*. Thousand Oaks, CA and London: Sage Publications, 1992.

23. U.S. Department of Health and Human Services, Archive for Recalls, Market Withdrawals & Safety Alerts, June 20, 2011. Available: http://www.fda.gov/Safety/Recalls/ArchiveRecalls/default.htm.

24. Center for Disease Control and Prevention, Multistate Outbreak of Salmonella Bredeney Infections Linked to Peanut Butter Manufactured By Sunland, Inc. (Final Update), October 12, 2012. Available: http://www.cdc.gov/salmonella/bredeney-09-12/index.html.

25. M. Nestle, *Pet Food Politics: The Chihuahua in the Coal Mine*. Berkeley, CA: U.C. Press, 2008.

26. G.R. Lawrence, Panoramic picture illustrating the beef industry, 1900. Available: http://www.loc.gov/item/2007663982/.

Index

Communication Practices in Engineering, Manufacturing, and Research for Food and Water Safety, First Edition.
Edited by David Wright.
© 2015 The Institute of Electrical and Electronics Engineers, Inc. Published 2015 by John Wiley & Sons, Inc.

Books in the
IEEE PRESS SERIES ON PROFESSIONAL ENGINEERING COMMUNICATION

Sponsored by IEEE Professional Communication Society

This series from IEEE's Professional Communication Society addresses professional communication elements, techniques, concerns, and issues. Created for engineers, technicians, academic administration/faculty, students, and technical communicators in related industries, this series meets a need for a targeted set of materials that focus on very real, daily, onsite communication needs. Using examples and expertise gleaned from engineers and their colleagues, this series aims to produce practical resources for today's professionals and pre-professionals.

Series Editor: Traci Nathans-Kelly

Information Overload: An International Challenge for Professional Engineers and Technical Communicators · Judith B. Strother, Jan M. Ulijn, and Zohra Fazal

Negotiating Cultures: Narrating Intercultural Engineering and Technical Communication · Han Yu and Gerald Savage

Slide Rules: Design, Build, and Archive Presentations in the Engineering and Technical Fields · Traci Nathans-Kelly and Christine G. Nicometo

A Scientific Approach to Writing for Engineers & Scientists · Robert E. Berger

Engineer Your Own Success: 7 Key Elements to Creating an Extraordinary Engineering Career · Anthony Fasano

International Virtual Teams: Engineering Global Success · Pam Estes Brewer

Communication Practices in Engineering, Manufacturing, and Research for Food and Water Safety · David Wright